Diagnostic Veterinary Ultrasonography

NIPA® GENX ELECTRONIC RESOURCES & SOLUTIONS P. LTD.
New Delhi-110 034

About the Authors

Dr. Ramesh Kumar Chandolia graduated from College of Veterinary Sciences, Haryana Agricultural University (H.A.U.), Hisar in 1982. He completed his M.V.Sc. degree from same university in year 1985. He obtained his Ph.D. degree from University of Saskatchewan, Saskatoon, Canada in year 1996. His academic service started in the year 1985 as assistant professor in the department of Veterinary Gynaecology & Obstetrics, H.A.U. Hisar and got promoted as Associate Professor in year 1998 and to the rank of professor in the year 2006. He served officer of the university as Professor and Head of Vet. Gynaec. & Obstetrics dept; Regional Director of RVDEC, Director (HRM) etc. He also served as Dean of several Veterinary college of the Country viz. SG Veterinary college, Sriganganagar; RPS college of Veterinary Sciences, Mahendergarh, MJF College of Veterinary Sciences, Chomu and Ramkumari Veterinary College, Mukundgarh. He was awarded DAAD fellowship in 1988 and then postdoctoral fellowship in the year 1997 at University of Florida, USA and then in 2005 in Germany. He has visited several countries viz. Australia, Finland, France, Germany, UK, USA. During his Ph.D. he got training on ultrasonography at University of Saskatchewan under eminent Professors viz. R.A. Pierson, N.C. Rawlings, R. Mappletoft, Greg Adams and many more. Dr. Chandolia initiated ultrasonography trainings in India on national level in the year 2004 and subsequently in 2008. He trained veterinarians of several states in India. He delivered lead papers on ultrasonography in ISSAR Conventions and International conferences. He was invited to delivered special lectures by other universities for trainings of veterinarians. The images given in the book are mostly from his ultrasound lab.

Dr. Prem Singh is B.V.Sc. & A.H and M.V.Sc. and Ph.D. in Veterinary Surgery and Radiology from Haryana Agricultural University, Hisar. He had his postdoc experience from UK. He served H.A.U Hisar and Lala Lajpat Rai University of Veterinary and Animal Sciences as Assistant, Associate and Professor. He had been dedicated clinician and got 17 awards from various institutions. He served as Director of TVCC, LUVAS. Dr. Prem Singh worked tirelessly to diagnose the clinical cases using ultrasonography. His students worked on various aspects of surgery using ultrasonography and produced good images provided in the current book. Presently, he is serving as Dean, IIVER, Bahuakbarpur, Rohtak.

Diagnostic Veterinary Ultrasonography Principles and Practices

R.K. Chandolia
B.V.Sc. & A.H., M.V.Sc., Ph.D. (Canada), Post doc (USA) DAADF (Germany)
Dean, MR College of Veterinary Sciences and Research Center
Jhajjar, Haryana
Ex-Professor & Head
Veterinary Gynaecology & Obstetrics
Lala Lajpat Rai University of Veterinary & Animal Sciences, Hisar, Haryana

Prem Singh
B.V.Sc. & A.H., M.V.Sc., Ph.D., Postdoc (UK)
Professor
IIVER, Bahuakbarpur, Rohtak, Haryana
Ex-Director, Veterinary Clinical Complex
Lala Lajpat Rai University of Veterinary & Animal Sciences
Hisar, Haryana

NIPA® GENX ELECTRONIC RESOURCES & SOLUTIONS P. LTD.
New Delhi-110 034

NIPA® GENX ELECTRONIC
RESOURCES & SOLUTIONS P. LTD.

101,103, Vikas Surya Plaza, CU Block
L.S.C. Market, Pitam Pura, New Delhi-110 034
Ph : +91-11-43860225, Mob.: +91 9717133558, 9540816132
E-mail: newindiapublishingagency@gmail.com
Website: www.nipaersources.com

Print ISBN: 978-93-5887-762-5

ebook ISBN: 978-93-5887-975-9

Composed and Designed by NIPA®.

Acknowledgement

The authors acknowledge the contributions of postgraduate students who work in our lab and contributed to the generation of some of the ultrasound images; these contributor students were: Dr. Gyan Singh, Dr. Kailash Kumar, Dr. Jinsa George, Dr. Pradeep SV, Dr. Suresh Bhatia, Dr. L. C. Ranga, Dr. Devender, Dr. Ajit Soni, Dr. Pinki, Dr. Sarita Yadav, Dr. Sandeep Kumar, Dr. Vinay Yadav, Dr. Ravi Dutt, Dr. Sonu Kumari, Dr. Neeraj Arora, Dr. Sandeep Saharan, Dr. Rajesh Sahu, Dr. Vishal Yadav, Dr. Jitendra, Dr. Ramphal, Dr. Nafis Assad, Dr. Praveen Rose and Dr. Shaym Sunder. Support from Dr. Jaswant Singh and Dr. R.K. Sharma, Dr. Phulia, Dr. S.K.Chawla, officers of EBS Hisar and chapter collaborators

Preface

This book "Diagnostic Veterinary Ultrasonography: Principles and Practices" is intended for the beginner-ultrasonographers in the veterinary and animal Sciences, particularly for those veterinarians who are involved in providing services to large and pet animals. In India, ultrasonography of animals is hampered by PC & PNDT act, a law that prohibit sex diagnosis in human. Under this act, there is strict regulation for the movement of ultrasound machine at farmers doorstep. This regulation also covers ultrasound imaging by veterinarians in India. The authors collected data from cases from animals presented for clinical diagnosis. Images were self-generated, using several machines viz portable and fixed ones having color Doppler and 3D facilities.

Basic principles of ultrasonography have been explained along with a special chapter on physics of ultrasonography.

The reader will learn where to place probe for obtaining images of various organs. Normal images of various organs and abnormal changes in diseases have also been provided for comparison. Age related developmental changes and dimensions of various organs have also been covered.

Various species viz cattle, buffalo, sheep, goat, bitch etc., have been covered. Images of different organs viz., kidney, liver, spleen, gall bladder, urinary bladder, reproductive organs have been shown. This will help the learner to go ahead with imaging of these organs and will get help in comparing their images with the images given in this text.

It is hoped that the text and ultrasonic images generated and provided in this text will help one and all who wishes to go ahead for ultrasonography of the animals.

Good wishes.

R. K. Chandolia
Prem Singh

Contents

Contributors' Acknowledgement

1. **Dr. Gurdial Singh,** Contributed text on "Anatomy of some of the organs in dog and horses in relation to ultrasonography".
2. **Dr. R.K. Sharma,** contributed chapter on "Reproductive ultrasonography in buffaloes".
3. **Dr. N. K. Khurana,** contributed text on "Basic ultrasonography for fieldveterinarians assessing bovine reproductive system".
4. **Dr. Akspreet,** contributed part of text on "Ultrasonography of the liver and the kidney in canine, and Ultrasonography of stomach in dog".
5. **Dr. L. C. Ranga,** contributed chapter on "Ultrasonography of testis and accessory sex glands in bull".
6. **Dr. S.K. Phulia,** contributed three chapters viz. "Ultrasonic characterization of follicles and corpus luteum"; "Ultrasonography of mammary glands in buffalo" and "Ultrasonography for diagnosis of mastitis and teat Abnormalities".
7. **Dr. Sumant Vyas,** contributed chapter on "Ultrasonography in camelreproduction".
8. **Dr. Rajesh Sahu and Dr. S.K. Chawla,** contributed chapter on "Ultrasonography of gastrointestinal tract in canine".
9. **Dr. Kailash Kumar,** contributed in chapter on "Ultrasonographic images of goat fetus in utero and its internal organs from day 24 onwards".
10. **Dr. Gyan Singh,** contributed chapter on ultrasound images from clinical cases and provided images used in the text.
11. **Dr. Jinsa George,** contributed chapter on "3D-ultrasound images taken sequentially for fetal development in dog".

1

Physics of Ultrasound

Ultrasound is important tool for diagnosis and research that can be compared with other tools like radio-immunoassays. This has become an important tool applicable to farm and pet animals. Ultrasound uses high-frequency sound waves to produce images of soft tissues and internal organs. Electric current is applied to crystals in the transducer, producing vibrations characteristic of the crystals and resulting in sound waves. The sound beams that pass through the tissues are quite thin and a thin "slice" of tissue is sampled. The two dimensional image produced by ultrasound resemble to histological section. The sound waves are propagated or reflected by various tissues in different ways depending on their thickness and composition. The transducer receives those reflections and converts into electric signals and produces an echo on the ultrasound screen. Grey - scale is the term used for monochrome monitors on which a number of grey shades are distinguishable between black and white and constitute the image of reflected portion of tissue. Non-ecogenic (non-echoic) tissues like fluids do not reflect sound waves and appear black on the screen. The echogenic tissues, having dense consistency, appear white as they reflect most portion of sound beam. Sometimes the artifacts create problem in the real image of the organ. These artifacts are created by certain tissues that cause bending or back and forth bouncing or blocking of sound beams. For imaging the reproductive tract of the domestic animals, B-mode (brightness modality), real time scanners are used. The amplitude of returning beams is reflected in the brightness of the dots on the screen. B-mode modulates brightness of the dots or pixels on the screen of echo displayer. The echo signals are presented as brightness of pixel corresponding to the amplitude of signal. The picture should be adjusted for clear display. In B-mode display the time-base brightness should be reduced just below the threshold of visibility so that even the smallest echoes could be visible. The contrast and bright should be adjusted on the monitor. Standard settings should be laid down for comparison to images from different dates. The important setting is of time gain compensation that issued to increase the amplification. Real-time is a term used to describe a process or system in which there is negligible time delay between the input of data and the output of processed data. This term is generally used for B-scan systems in which the scanning is performed rapidly by electronic or electromechanical means, such as the use of multiple transducer arrays. M-mode (motion mode) of imaging has been incorporated in B-mode scanners. Doppler ultrasound system can be used to monitor fetal heart beat and blood flow in large vessels in late pregnancy or other body organs. The A-mode (amplitude modality) was used earlier for the diagnosis of pregnancy in sheep and mare. A-mode scanners produces one-dimensional display of echo amplitudes for various depths, therefore, now it is being used to measure the fat and lean composition of meat. Time base is the line representing the time-axis upon the display screen. The time base may be visible, as in the case of an A-mode display, or suppressed, as in the B-mode case. In modern age the digitization of echo signal has achieved by using fast digital circuits and computational speed to operate on the megahertz frequency signals and maintain the amplitude precision.

Acoustic Physics

To have a good image, understanding of acoustic physics and technical terms is important. Attenuation is a decrease in amplitude and intensity of sound beam. It is caused by scattering and reflection of acoustic energy at interfaces and absorption of the beam. Attenuation is considered important because it limits the depth range in tissue that can be imaged, especially with high-frequency sound beams. It also helps in diagnosis as the relative attenuation of a region can be qualitatively estimated in comparison with adjacent tissue. The relative rate of attenuation of an ultrasound beam depends on both the tissue traversed and on the ultrasound frequency. Attenuation increases with increase in frequency. The attenuation rate for a 5-MHz beam is 5 times the rate at 1 MHz. The attenuation coefficient is given in decibels per centimeter. Attenuation coefficient is considered as convenient way to express attenuation because it eases in calculating signal decrease for a particular distance traversed.

Net attenuation = a x d: where a is the attenuation coefficient in decibels per centimeter and d the distance traveled.

Absorption is the loss of energy from an ultrasonic wave during propagation by conversion into another form usually heat. This is one mechanism of attenuation.

Damping is a technique, involving electrical or mechanical loading, used to reduce the duration (together with amplitude) of an ultrasonic pulse generated by a transducer, and/or of echo pulses received by the transducer.

Decibel (dB) scale: A logarithmic scale for describing the intensity of a signal in terms of reference intensity.

Relative intensity (dB) = 10 log (I1 /I2) : where I1 and I2 are two intensities. The intensity is proportional to the square of the amplitude.

Delay-line: A device for delaying an input signal by a precise period of time; useful for calibrating a caliper.

Diffraction is the tendency of waves to bend around obstacles which are small in size compared with the wavelength.

Doppler Effect is the effect by which the frequency of an observed wave is changed with respect to its frequency at source, due to relative motion between source and observer.

Frequency is the number of complete cycles occurring in second. It is the reciprocal of the period, and is expressed in hertz (1 Hz = 1 cycle per second). The frequency ascribed to a pulse-echo transducer is only a nominal, average figure.

Gain is a measure of the amplification of a device or system and is defined as the ratio of the amplitude of a variable at the output of the device or system to that at its input. It may be expressed in decibel (dB), the input amplitude serving as the reference.

Modulation is the process of superimposing information upon a continuous wave by varying, with time, a parameter such as amplitude or frequency. The train of short, oscillatory echo- signals, received by the transducer as a result of severe amplitude-modulation of a continuous wave.

Resolution is the minimum spatial separation of two point targets at which separate registrations can just be distinguished on the display.

Scanning frame is a mechanical arrangement for confining the movement of the probe to the desired plane, and including sensors for monitoring the spatial position of the transducer and the direction of its axis.

Time gain compensation is the process by which the gain of a pulse-echo system is varied with time, following the generation of the transmitter pulse, to compensate for the attenuation by proximal tissues of echoes originating from deep-lying targets.

Echogenic: Echo-producing - structure or medium.

Gray - scale is the term utilized for monochrome monitors on which a number of grey shades are distinguishable between black and while.

Linear is the term used to describe the relationship between two variables which is arithmetically proportional, the rate of change of the second variable remaining constant as the first variable is changed.

The piezoelectric properties of certain crystals are being utilized in the modern ultrasonography. The piezoelectric properties of the crystal is such that it can both generate sound waves when charged with an electric signal and generate an electric signal when struck by the returning sound waves. The delay between propagation of the wave and reception of the echo is used to determine the distance from the crystal to the reflector. These piezo-crystals expand and contract, depending upon alternating polarity of the electric signal, when electric current is passed and produce an electric current, when compressed by returning echoes. The expansion of piezo-crystal causes compression of neighboring tissue molecules, and contraction causes rarefaction of these tissue molecules. A beam is the sound field resulting from firing a cluster of elements. A beam involves one pulse and time is allowed after the firing of each pulse for the collection of echo information. For each beam, the echo pattern is collected in a fraction of a millisecond because velocity of the ultrasound pulse in tissue is 1.54 mm/msec (1540 meter/sec). Image is formed when a sweep of beams across all elements is completed. Generally, the frame rates are 20 to 30 per second, images of rapid movement like heart beat or wandering of equine embryo can be observed with this rate but faster movements requires higher frame rate.

Pressure amplitude and acoustic intensity are variables to make quantitative statement about strength of a sound wave. Acoustic intensity (I) is related to square of pressure amplitude (P0).

I = Square (P0)/2 rC :where r is the density of the medium and C is the speed.

Transducer Arrays

Transducers produces piezoelectric effect i.e. they expand and contract when an electric signal is applied, producing a sound wave in the medium. Every transducer has a resonance frequency at which it operates most. The thickness of piezoelectric element determines the frequency of transducer. Thicker the element, lower is the frequency of transducer. The quartz, naturally, has piezoelectric property and was used in earlier transducers and now is replaced with piezoelectric ceramics, such as lead zirconate titanate or lead metaniobate. These materials are found to have to have shorter pulse duration and more sensitivity. Modern transducers are composites having grid arrangement with the piezoelectric material sticking out like posts. A type of epoxy is used to fill the area between the posts. These modern transducers provide large frequency bandwidths and acoustically are more similar to soft tissues. These ceramic materials are polarized to work in transducers. This is done by heating the element by applying a strong electric voltage, and cooling (see Bushong and Archer, 1991; Zagzebski, 1994). Therefore, it has been advised not to sterilize the transducer by heat. The good quality transducer is judged by spatial resolution that refers to how closely positioned two reflectors or scattering regions can be seen as separate reflectors on a display. Another property of good imaging involves pulse duration and axial resolution. Axial resolution is more important as it refers to how closely positioned two reflecting targets can be along the sound beam axis and still be distinguished on the display. It depends on the duration of the each ultrasound pulse emitted by transducer. There is relationship between damping, pulse duration and axial resolution. The greater is the amount of damping, the shorter will be pulse duration and better will be axial resolution. To have a distinct image of two closely spaced reflectors the effective-beam width is also important.

The active surface of transducer is easily identified and it is advised to protect from any electric shock. The coaxial cable connects the transducer to the console. The change of transducer from 5 MHz to 7.5 MHz, the attenuation rate will increase to 2.5 dB/cm. This will decrease the depth of compensation area by increasing frequency. Along with correct type of transducer proper control of distal echoes is important too. The decrease of overall sensitivity by 10 dB will decrease amplitude of all displayed echo, including deeper tissues by 10 dB (see Zagzebski, 1994).

The ultrasound pulse distinguishes between two closely spaced reflectors and this is called resolution. If the reflectors are located at right angles to the direction of reflectors then we get lateral resolution and if the reflectors are located along the direction of the beam, it is called axial resolution. High-frequency transducers give better axial resolution due to shorter pulses. These short pulses are less likely to overlap two neighboring reflecting surfaces, and distinguish them better. The axial and lateral resolutions with 3 and 5 MHz transducers might be 1, 3 mm and 0.5 and 2 mm, respectively. The use of focusing can affect the image as focusing narrows a portion of the beam profile and thus increases the amplitude of echoes from the reflectors at a certain depth. If the beams are focused in the thickness plane, it gives better lateral resolution. Focusing of different transducers could be different. Higher frequency transducers have focal region closer to the transducer. The comparable focal distances of different transducers, used in animal science are 70, 35 and 20 mm for 3.5, 5 and 7.5 MHz transducers. The effective frequency for high-quality imaging is equal to two times the focal distance. Focusing at particular point can influence the image quality. In some transducers there is automatic dynamic focusing while in some machine it is switch-selectable. Using these dynamic focusing variable focal distances can be used rather than fixed distances (see Kremkau, 1984).

Field

Field is the ultrasonic region in which an ultrasonic wave propagates. In some transducers the field may be confined to a beam, as a result of interference phenomena.

Near field is the region of the ultrasonic beam, approximately cylindrical region of beam, immediately in front of the transducer. Within this zone there are significant, complex intensity variations, with hot and cold spot resulting from interference effects. Near field is generally the first half and far field is the distal half. The junction of near and far beam, in an unfocussed beam is at the point of natural divergence of the beam. The resolution is better in the near field as beam starts diverging in the far field. Far field is the region of ultrasound beam beyond near field, in which the intensity decays smoothly with distance (Pierson, 1988; Ginther, 1995). This zone consists of a divergent main lobe and a number of relatively weak side lobes.

The near gain is important in those images where tissue is closer to the transducer and those cases the far gain is not that important. Proper balancing of controls provides better clarity of image. Far gain is important when visualizing large or distant structures. The near gain controls initial attenuation. This sets the amplifier gain from starting to maximum. Setting the near gain at a particular value describes that particular value in decibels below the maximum. The effective beam is confined to a region close to the axis of the transducer. Extension of near field is outward from a transducer and is given by:

Near Field length = square of radius of transducer/l : (l is the wavelength).

In the far field the beam gradually weakens and diverges. The beam width, W, at axial distance, z, in the far field of an unfocussed transducer may be approximated by:

$W = 1.22\ \lambda z/2a$

The beam width decreases with decreasing the wave length (l). Wave length is inversely proportional to frequency and beam width decreases with increasing frequency. In the far field, the beam width decreases with increase in size (radius) of transducer (Zagzebski, 1994). It is also important to note that length of near field also increases with larger diameter transducers and beam in the far field are less divergent. Another important observation about near field is that its length increases with increase in frequency of transducer.

Propagation of Sound Waves

Zagzebski (1994) has nicely described propagation of sound waves while reviewing instrumentation of diagnostic ultrasound. His important considerations are being summarized below.

The ultrasound transducer is used to transmit high frequency sound waves and echoes from tissue interfaces are detected and displayed. Continuous waves are produced when transducer oscillates continuously. The continuous beam is not commonly used in diagnostic ultrasound rather pulse mode is used where transducer, on activation, vibrates for few cycles. During such action an ultrasonic pulse is propagated through tissue.

The longitudinal waves are produced by transducers because such type of sound waves can propagate through soft tissue. The speed of sound though tissues is also important. The average speed of sound is between 1500 - 1600 m/second in most of soft tissues. Reviewers (Fish, 1990; Zagzebski, 1994) are also describing importance of frequency of sound waves. Frequency depends upon vibrating source and defined as oscillations per second determining quantity for relative amplitudes and intensities of transmitted or reflected waves. The difference between acoustic impedance on either side of the interface determines amplitude of reflected wave.

The ratio of the reflected to the incident pressure amplitude (R) is determined by the acoustic impedance and its unit is *hertz* (Hz). The sound waves greater than 20 MHz are referred as ultrasonic. Generally, ultrasound frequency fall in the range of 2 to 15 MHz. Period of wave is inverse of frequency (Period = 1/frequency). Wave length has been referred as a distance over which wave variable like pressure; cyclically repeat their values at any instant of time. Partial scattering and reflection of a sound beam at tissue interface is also important. Two differing tissues form interface. Acoustic impedance is on the distal side of the interface (Z2) and the impedance on the proximal side (Z1).

$R = (Z2 - Z1) / (Z2 + Z1)$.

It is important for the ultrasound equipment to pick up the small reflected echoes because some time the impedances of the two tissues forming interface are nearly the same and the ratio of the reflected to the incident amplitude is small. It is fortunate that most of energy of sound beam gets transmitted through the interface.

Specular reflectors are large smooth interfaces. Generally the same transducer is used as both source and detector in ultrasound examination oblique incidence of sound beam results in a weak echo signal or no echo signal detected from a specular reflector. Perpendicular beam provides strongest echo. Oblique incidence at an interface could result into refraction, a deviation of the transmitted beam across the interface. Zagzebski (1994) has cautioned to be aware of refraction phenomenon as it could result into artifacts appearing on image. There are more chances of refraction artifacts in those interfaces where one of the tissues is fat.

Transducer detects scattered echoes to produce grey-scale textured appearance of structured. Scattering depends on the ultrasonic frequency. Change in frequency may change the scattering and that may alter image quality, particularly in B-mode scanning. Subtle changes in the brightness level and texture of image are considered due to regional variation in the level of ultrasound scattering (Ginther, 1995).

Velocity of sound waves in body tissues:

Sr. No.	Tissue	Velocity (m/sec)	Acoustic impedance	Descending order of echoicity	Frequency and its Wavelength
1	*Bone*	*4080*	7.8	Bone, gas	10MHz; 0.15 mm
2	*Lens of eye*	*1620*	1.84	Fat, vessel walls	7.5 MHz; 0.21 mm
3	*Muscle*	*1585*	1.70	Kidney sinus	5.0MHz ; 0.31 mm
4	*Blood*	*1570*	1.61	Prostate Gland	3.0 MHz; 0.51mm
5	*Kidney*	*1561*	1.62	Spleen	2.0MHz; 0.77 mm
6	*Liver*	*1549*	1.65	Liver	*2.0-3.0 MHz; 0.77 & 0.51 mm*
7	*Soft tissue*	*1541*	1.58 (Brain)	Kidney cortex	5.0MHz ; 0.31 mm
8	*Water*	*1540*	1.54	Muscles	5.0MHz ; 0.31 mm
9	*Fat*	*1450*	1.38	Kidney medulla	5.0MHz ; 0.31 mm
10	*Air*	*331*	0.0004	Bile in Gall bladder	5.0MHz ; 0.31 mm

Focused Vs Unfocused Beam

To achieve better resolution many transducers are focused. Focusing can be done by using acoustic lens or using a curved piezoelectric element. This creates a focal zone but reduces beam width. This type of beam is wider near transducer. The intensity increases but beam narrows down near focal distance. The beam diverges beyond focal point. Kossoff (1979) has provided the following calculation for beam width (W) for spherically curved transducers:

$W = 1.22\ \lambda F/D$ where D is the diameter of the transducer, F is the focal length and λ is wave length.

The emerging sound beam is the sum of individual beam from individually excited elements. The array arrangement has advantage as the focal distance can be adjusted simply by selecting a different delay sequence for pulses applied to individual elements. This can be demonstrated by one simple example; if initial attenuation is 50 dB and 2 cm of delay, then decay of echoes in up to 2 cm will be at regular rate as no compensation is applied in this region and then compensation will apply till certain distance. It has been seen in such cases, that total compensation is 2 cm more than usual. In many array transducers there is choice of transmit focal distance. This type of array system also permits focusing of echo signals. When echo signals are picked up by elements in array, signals from individual elements are added together in the receiver. There is slight disadvantage, as echo from a discrete target located in the beam will arrive at individual elements of the array at slightly different times because of slight differences in distance. This leads to weak signal. This problem can be overcome by designing a proper delay so that precise synchronization can be done before summing. Dynamic focusing can be helpful for sharp image. The initial focusing is set at a shallow distance, as the time after the transmitted pulse increases, the receiver focus changes automatically.

Mode of Display

The investigators have described 3 modes of display in echo signaling:

A mode shows echo amplitude versus distance and is used in ophthalmology and to measure fat thickness in animals.

B-mode is more commonly used. It converts echo signals into intensity-modulated dots on the display. In this mode imaging is performed by sweeping the axis of the ultrasound beam over the volume of organ and display is obtained. In this mode 100 to 200 separate lines of beam are involved in image acquisition. These different lines from slightly different directions are positioned in a location on screen corresponding to the reflector position. The buildup of digital memory, in this mode, is stored in many machines this process is also called scan converter. The vector scan format is accepted by scan converter and store these signals in a location in memory. The signal is passed through an analog-to digital (A/D) converter before storage. The memory of scan converter is made of matrix of individual addressable cells. Individual cells are referred to as picture elements or PIXELS. Further, each pixel is broken down into binary storage units called bits. The number of amplitude levels or shades of gray that can be stored in each memory location can be calculated by 2n (n is the number of bits at each pixel location e.g. 5 bits allows 32 levels of gray, 6 bits allows 64 and 8 bits allows 256. To produce large format, 500 x 500 pixel matrices is useful. To represent the echo amplitude in each pixel, at least 8 bits should be available (see Goldstein, 1991).

M-mode is also called motion mode and image is obtained by slowly sweeping a B-mode trace across a screen. It is used to obtained movement pattern of walls and valves in echocardiograph, to observe heart-beat of fetus and movement of embryo in some species.

Dynamic Range

The capacity of equipment to handle echo signals is called dynamic range (Shirley, 1978). The input amplifier of equipment has higher dynamic range than display stage. The echo compression is applied in the receiver so that a wide variation in the echo signal amplitude can be applied. In B-mode the signal pre- processing signal is passed to distal value in memory and then processing is done and post processed signal is displayed.

It has been described that while imaging in pulsed-echo system, a short duration burst of ultrasonic energy is transmitted into body and echoes from scattered and interfaces are detected and displayed. The distance between transducer and reflector can be calculated from pulse-echo equation as given below:

Distance = 1/2ct where c is the speed of sound in the medium and t is the time interval between when the pulse was emitted and when the echo signal was picked up. The average speed of sound in soft tissue is 1540 m/second. If the 't' is 13 microsecond it corresponds to a reflector depth of 1 cm (Zagzebski, 1994).

Safety and Limitation of Ultrasound

There is growing concern about safety of ultrasound (see Blackwell, 1988; Kremkau, 1991 and Tarantal and O'Brien, 1994). Two main points, in this regard, were highlighted. The interaction of ultrasound with tissue leading to production of heat is one important aspect to be taken care of. Exposure to ultrasound for more than one hour can elevate the temperature between 2.5 and 5°C. (Miller and Ziski, 1989: cited by Tarantal and O'Brien). This hyperthermia may lead to teratogenic effects and may affect nervous system (see Warkany, 1986). However, it has been suggested that degree of hyperthermia depends upon intensity of ultrasound. Other complication of ultrasound could be cavitation or gaseous bubble formation. The mechanism of cavitation has not been well documented in mammalian reproduction.

In addition to above safety problems, other limitations of ultrasound imaging measurements have been reviewed by Burns *et al* (1992). The acoustic impedance which is product of density and speed of sound in the tissue is influenced by tissue density, compressibility and temperature. The interfaces between soft tissues transmit most of ultrasound pulse and only a very small portion of the pulse will be reflected back. For this reason, many tissues composed of different tissues but appear identical. The arising of echoes from soft tissues and fluids is result of Rayleigh scattering. Some weak echoes may return back and influence local maxima and minima pattern of ultrasound echoes. On one hand it is known that the speed of sound varies in different soft tissues but most of the time the speed is assumed to be constant, 1540 m/sec. It is generally considered that all echoes received by transducer arise from a 90° angle but many structure can deviate the echoes in the direction of ultrasound beam can be recorded and can affect ultrasound image. Another important drawback of image geometry is fundamental limit of precision of a measurement made from an ultrasound image, because the image is constituted by series of cross sections of a slice of finite thickness. Axial resolution is the minimum separation of two targets that lie parallel to the direction of sound beams. This resolution is affected by production of acoustic pulse which has extended length in space. The lateral resolution is the separation of two targets located at 90° angle to sound beams. There are limitations to record any measurements particularly in highly focused beam. This type of beam can achieve good lateral resolution in the focal zone but poor in the near and far filled regions. This affects the measurements that may vary with depth of tissue, degree of focusing and size of transducer. The thickness of slice is derived from spherical surface but the problem arises for curving surfaces e.g. testis where edges are fuzzy and affect measurements.

Another problem discussed by these reviewers is the contrast resolution, that depends on tissue composition and amplitude of beam. This presents problem for ultrasound frequency. To have better spatial resolution the frequency should be higher but it should be lower to improve signal amplitude. Therefore, use of an optimum frequency is always a compromise.

Other errors, discussed by Burns *et al* (1992) come from faulty scanning technique and problems of interpretations. The main scanning errors could be failure of selection and alignment of correct scan plane, failure to confirm initial measurements and wrong placement of clipper. There are three errors of interpretation 1) Failure to identify anatomy of normal and pathological tissue, 2) Failure to correctly find edges and 3) failure in preparation and comparison with reference data tables.

It has been suggested that reporting of data should be confirmed by others and accuracy of measurements should be mentioned.

Correction of Errors and Improvement of Scan Technique

Several corrective measures for errors for improvement of examination have been described (Shirley *et al.*, 1978). Following are the main suggestions.

Preparation of animal or particular area: The transducer should be in close proximity of organ therefore, clipping of hairs, removal of bowel and application lubricant are some of important point to be taken care of. Different coupling media are not reported different therefore either arachis oil or paraffin oil or commercial gel can be applied. There are many conditions that may affect the image quality like fat around the organ, any scar or injury to the organ of interest.

Position of transducer and pressure applied: The transducer should be hand held and coordination of hand and eye need practice and should be to achieve better image. The pressure should be optimum because excessive pressure may distort the anatomy of the organ and less pressure may lead to loss of contact.

Movements of transducer: Linear and sector scanning or their combination is commonly used and movements are necessary to obtain right sized image of organ under ultrasound. With arc scanning, most of information is lost. The speed of movement of transducer should neither be too fast nor too slow. The fast speeding movement may lead to loss of echoes and too slow speed may lead to overwriting in some systems.

Homogeneity of image by evenness of scanning: This is important if inference depends on distribution of echoes. In B-scanners, the echoes from boundaries are specular and are brightest when sound beams strikes normally. The real time scanners view its field with even rate but orientation of scanning plane must be right. The size of transducer should be just right because larger scanner presents difficulty in achieving the right plane. All the slices of organ should be taken and examined before selection of final one. Labeling should be properly identified.

Transonicity: To obtain the far wall of structure the composition of tissue should be known because far wall of cystic structure will be brightened up as there is minimum attenuation, while solid structure attenuates the most and creates difficulty in obtaining the image of far wall.

Internal echoes: To study the detail of internal echoes precaution should be taken in grey-scale imaging because small echoes particularly due to artifact can disturb the cyst characteristics. If any abnormal structure is seen it is better to repeat and compare with known reference.

High Quality Image

Many reviewers (Maslak, 1985; McDicken, 1991b) have highlighted the importance of image quality and three important parameters are suggested for good quality image 1) detailed resolution 2) contrast resolution 3) uniformity throughout the field of view.

The detailed resolution has been further defined as the ability to distinguish small structure with clarity. The minimization of dot size is important for optimizing detail resolution e.g. 20 dB dot size = 3R/N where N is the array of active elements and R is the focusing distance under uniform intensity and one- half wavelength spacing.

The judgment of contrast resolution comes from differentiation of tissue type and to view subtle structures in the presence of very bright reflectors. Apodization is a spatial filtering function that optimizes contrast resolution by reducing side lobes. Apodization is available in some electronically scanned array systems. Uniformity of image is important to compare details and contrast resolution throughout the field view. Uniformity of image is achieved by Tracking-Lens System (TLS). This optimizes image detail and contrast resolution. TLS takes speed of sound and depth into consideration by using the timing information to focus the system at the specific point from which echoes are being received. There are some tools available to measure these three parameters of image quality. These tools are: Phantoms, beam plots and beam profiles. Phantoms are commonly used (Spaulding, 1992) while other two are used by researchers.

2

Introduction to Ultrasonographic Imaging Process

Ultrasonography is the process of obtaining images of an area, organ or tissue with the help of ultrasound beam that can be displayed on a screen. Such images can be recorded on videocassettes or printed on a thermal paper. Ultrasonography in veterinary practice is a revolutionary step for diagnosis of various ailments and to assess normal physiological functions of the body organs of domestic animals. Ultrasonography is considered a wonderful tool particularly for the study of reproductive functions viz. follicular dynamics, pregnancy diagnosis and abnormalities of contents of tubular tract. In 1980, Palmer and Driancourt reported for the first time, real time ultrasonography for diagnosis of pregnancy in the mare. This report generated considerable interest among veterinarians and scientists about ultrasonography and this new discovery contributed significantly towards understanding of functioning of the uterus and ovaries of mare (Chevalier and Palmer, 1982; Ginther, 1983; Ginther and Pierson, 1984a & 1984b), cows (Pierson and Ginther, 1988, Reeves *et al.*, 1984) and other species (Buckrell, 1988; Driancourt, 1991). The ultrasonography is fast replacing the age-old procedure of rectal palpation of genital tract for pregnancy diagnosis (Pieterse *et al.*, 1990). This technique has been used and advocated for the use in the academics and field practice.

Type of transducers: In veterinary practice both kinds of transducers i.e. linear array and sector type are being used. The linear array transducers might be simply flat linear or curvilinear. The simple linear transducer is good for the images that are close to the active surface of the transducer. This kind of transducer has limited length and width i.e. it gives dimensions or measurements equal to its own length and breadth. On the other hand the curvilinear transducers are used externally or trans-abdominally. This gives pie- shaped image and covers more area than simply linear transducer. The frequency of the transducers might vary, considerably. In veterinary practice, generally the transducers have frequency of 3.0 to 7.5 MHz. The 7.5 MHz is used for viewing smaller structures or details of a larger structure. 3.0 or 3.5 MHz frequency is used for larger structures. The most common frequency used in cattle and mare is 5.0

MHz, while in small ruminants 6.0 or 7.5 MHz is commonly used. Imaging depth of ultrasonography decreases with increase in its frequency. The imaging depth of 3.5 MHz frequency is 17 cm; 5.0 MHz is 12cm, while that of 7.5 MHz is 8 cm. The higher frequency in veterinary practice (6.0/7.5 MHz) is used for neonatal studies, early pregnancy diagnosis and images of ovaries of sheep and goat. The higher frequencies (3.0/3.5 MHz) are used during advanced stages of pregnancies to obtain images of uterus and its contents. There is better axial resolution of 7.5 MHz frequency (0.2mm) than either 5.0 MHz (0.31mm) or 3.5 MHz (0.44mm). For common understanding, it is suggested to use higher frequencies for smaller animals (including lab animals) and lower frequencies for larger animals. Another important suggestion given by experience sonographers is that higher the frequency, better is the resolution, but poor is the penetration capabilities and vice-versa. A sonographer has to select kind of transducer, required for a particular structure. These days, transducers of switchable (multiple) frequencies are available that make the job easy to change the frequency as and when desired to view a particular structure. The decision of selecting either a linear or sector transducers is also sonographer´s choice for a particular image. It is to be remembered that the dimensions of image with linear array transducer are equal to the dimensions of the transducers, while in case of sector transducer, there is more penetration capability and cover the area in an angular design.

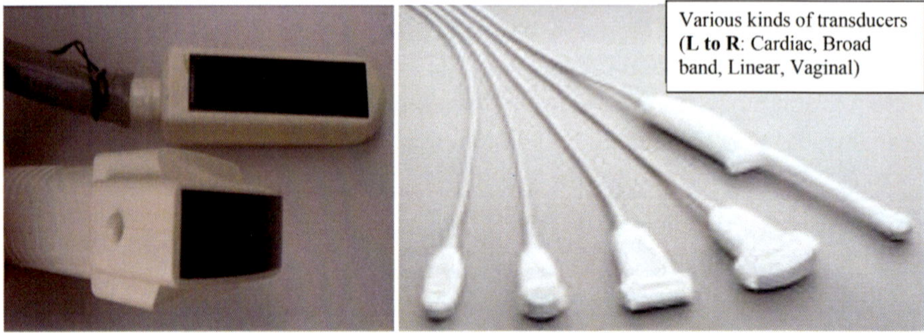

Photograph 1: Transducers/Probes

1. Rectal Linear & Convex probe

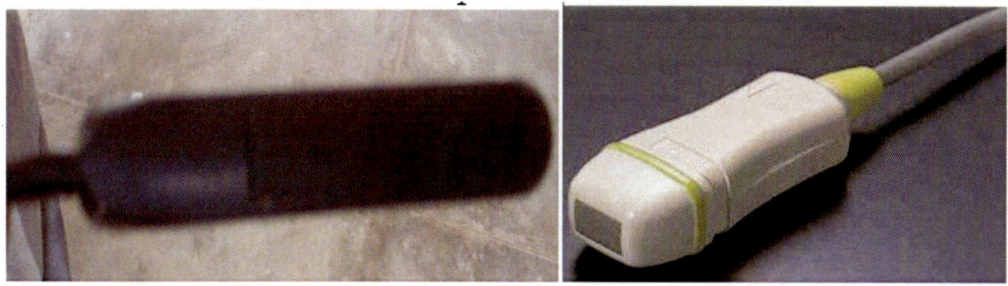

Rectal Linear

Cardiac probe

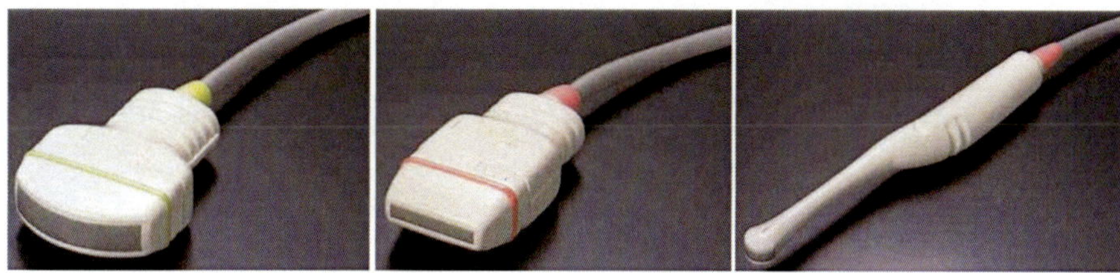

Convex probe

Medical Linear probe

Transvaginal probe

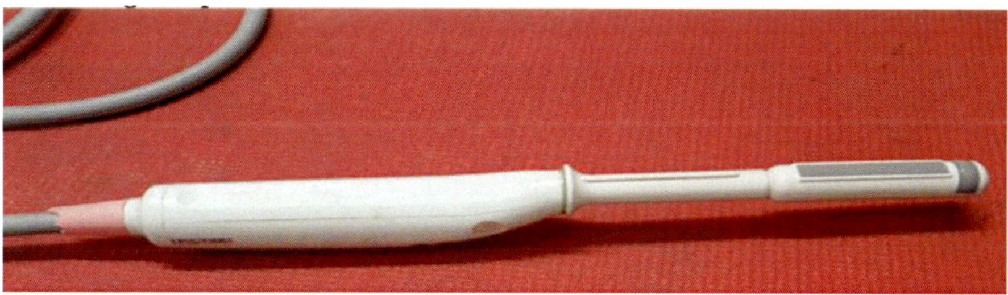

A probe with double scanner: Linear and circular

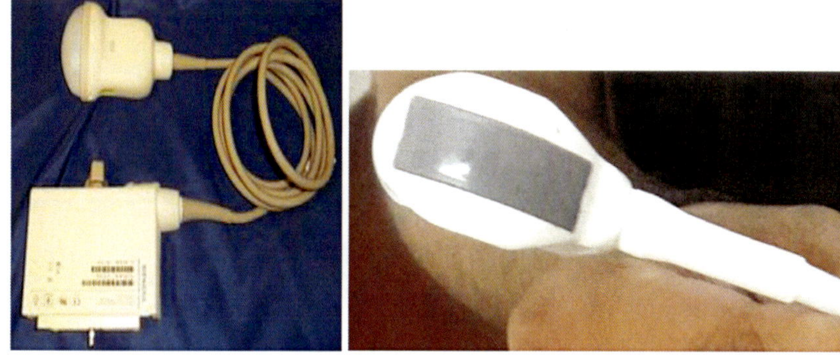

4D volumetric probe (left) and intra-operative probe (right)

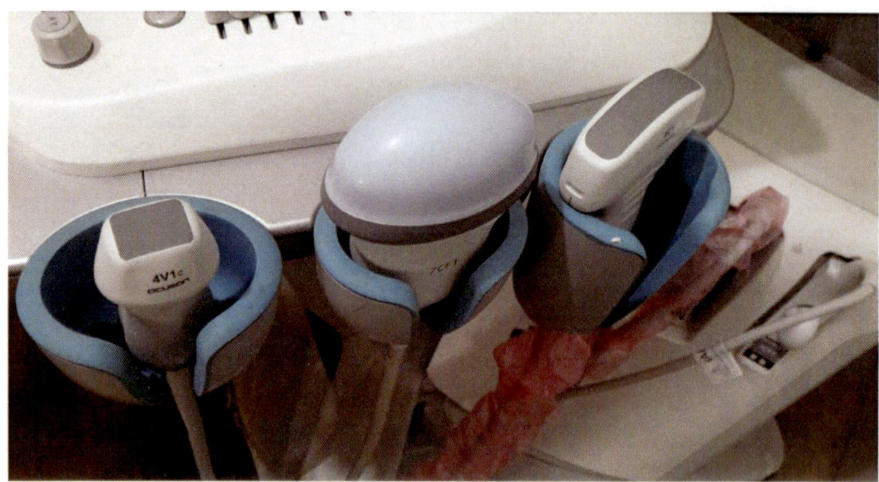

Various other probes viz. cardiac (left), 4D-volumetric (middle) and convex (right) probes.

Type of ultrasound machine

In Indian market many kinds of ultrasound machines are available. Many institutions are using a machine "Vet scanner 200" from Pie-medical (The Netherlands). Other machines used are from Toshiba (Japan), Aloka-Hitachi (Japan), Siemen (Germany), Philips, Sonocyte (USA), Madison (South Korea), and several other small machines from UK (France (ECM), Poland (Draminski) etc

Ultrasound Machines

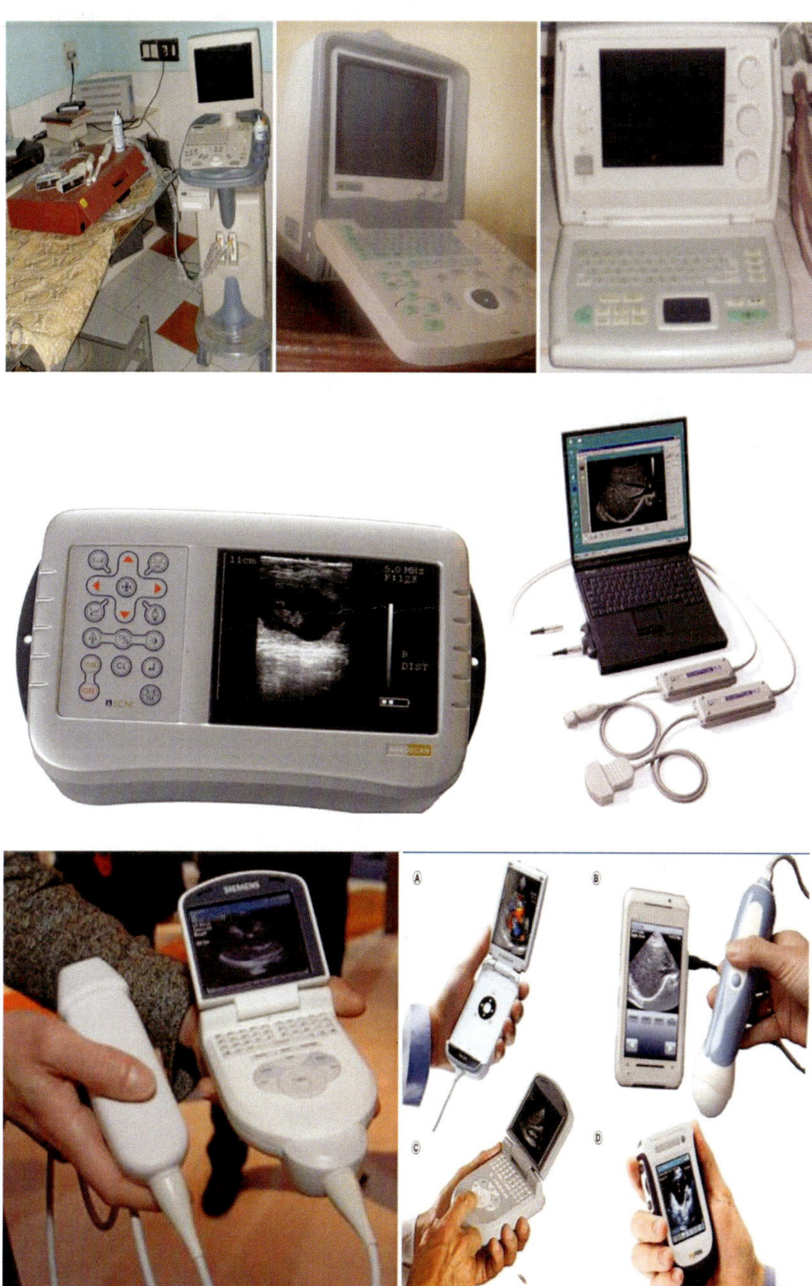

Photograph 2: Is showing few examples of large to small portable and battery- operated pocket ultrasound machines. Many mores are also available.

Ultrasound machine can generate black and white image and marked with B/W mark. One should be careful in purchase of such USG machines which are cheap. It is better to purchase machine that generate grey scale images. Grey scale images are better than black and white images. For comparative better view one should buy machine with Tissue Hormonic Image (THI) function. Image generated with THI are better than grey scale images. These days 3D and 4D machines with advanced features are available.

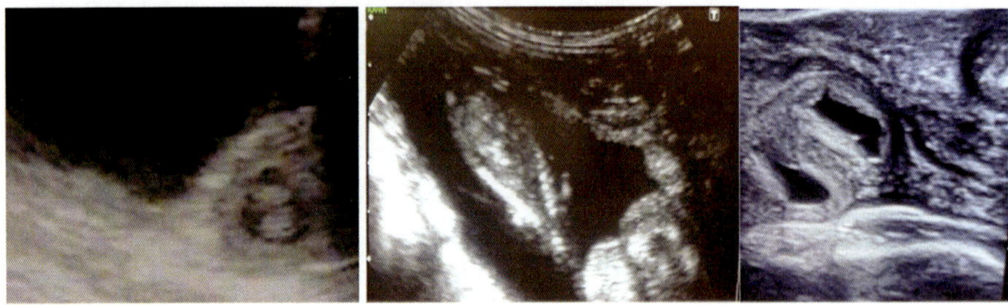

Fig. 1-3: are showing B/W (left), grey-scale (middle) and THI (right) image.

Goggle for Viewing Ultrasound Images

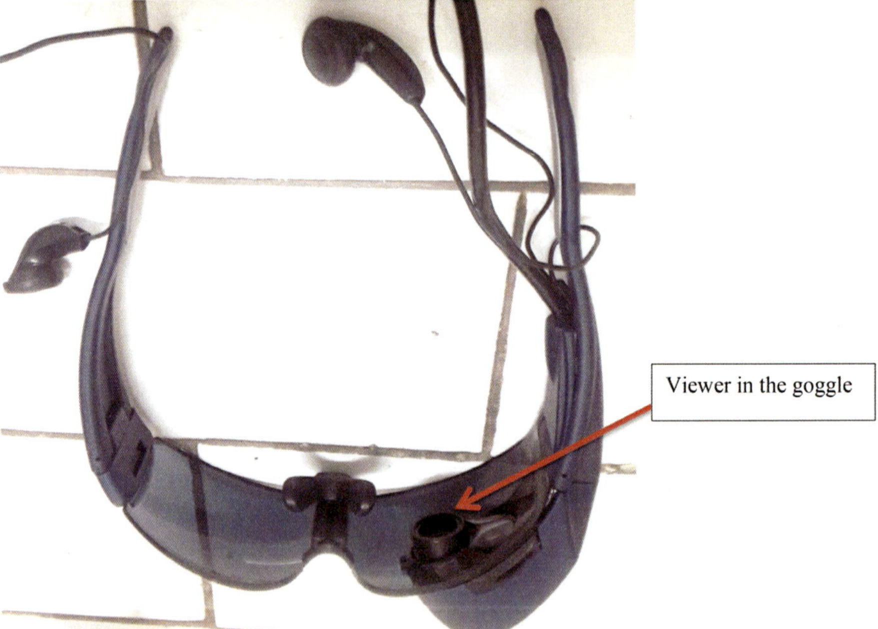

Photograph 3 is showing a goggle with USG-image viewer

Types of ultrasonography

In veterinary practice, four types of ultrasonography are in use viz. A-mode, B- mode, M-mode and Doppler ultrasound.

A-Mode: This kind of ultrasonography is used to know distance of a particular point in the form of amplitude of returned sound beam. This is also known as amplitude modality and is one-dimensional display of echo amplitude as per the distance or time. Generally, it is displayed in the form of graph, which provides distance on horizontal axis and amplitude on the vertical axis.

B-mode: It is brightness mode, which provides two-dimensional image of the tissue/organ scanned. It is most commonly used in veterinary practice as it provides graphic image on the screen. Brightness of the tissue/organ is in proportion to the amplitude of the reflected echoes returning to the transducer. Solid structures appear bright (white), while fluid appears black. The image appears live due to continuous updating of the image. Therefore, it is also termed as real time ultrasonography.

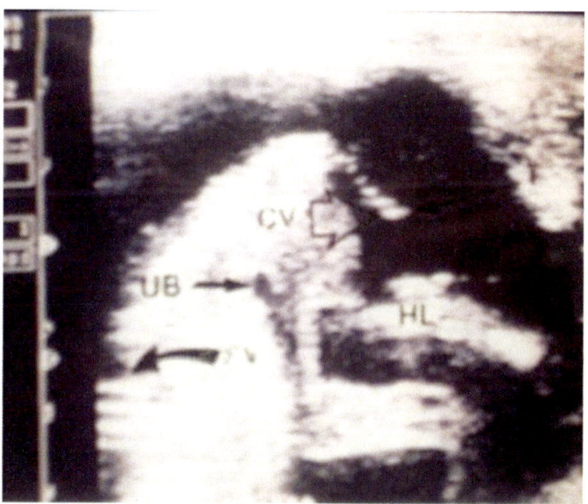

Fig. 4: is showing B-mode image from anechoic (black) to hyperechoic (white) stage.

M-mode: It is the motion modality, which provides tracing on the screen that can be recorded on a thermal paper. It is used to scan heartbeats or moving structures/tissues. It provides one-dimensional image on vertical axis versus time on the horizontal axis.

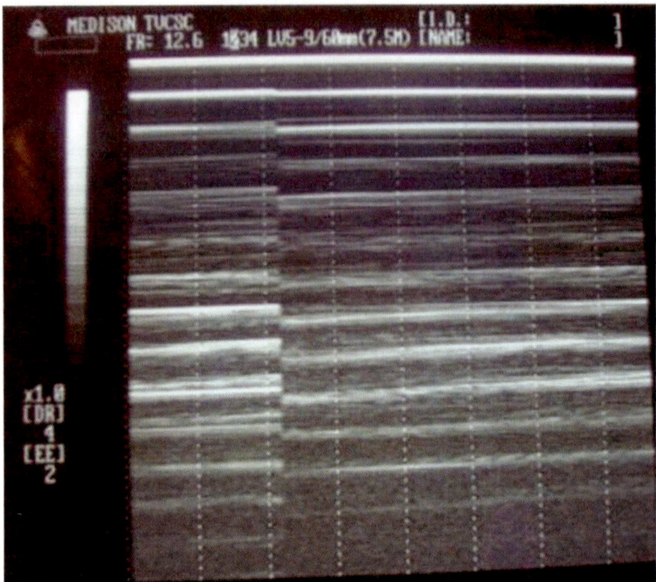

Fig. 5: is showing M-mode, but without movement with linear transducer

Combination of B-mode and M-mode is better than M-mode alone as shown below.

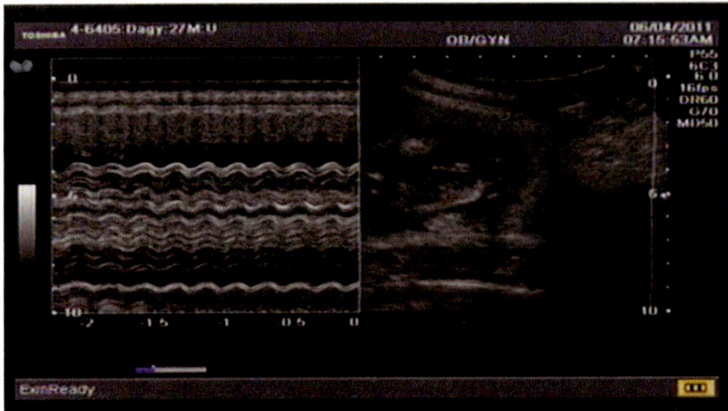

Fig. 6: is showing M mode on left panel and B-mode in right panel

Doppler ultrasound: It also works on ultrasound-beam reflections and is used for moving objects. If the motion of the moving object is towards the transducer, then the frequency of the sound beam is increased. If the motion is away from the transducer, then the frequency of the sound beam is decreased. It is of two kinds: 1) Continuous-wave Doppler system and 2) Pulse-wave Doppler system. In continuous-wave Doppler system, there are separate transmitting and receiving crystals, while in pulse Doppler; there is a single crystal for doing both the jobs (transmission and reception).

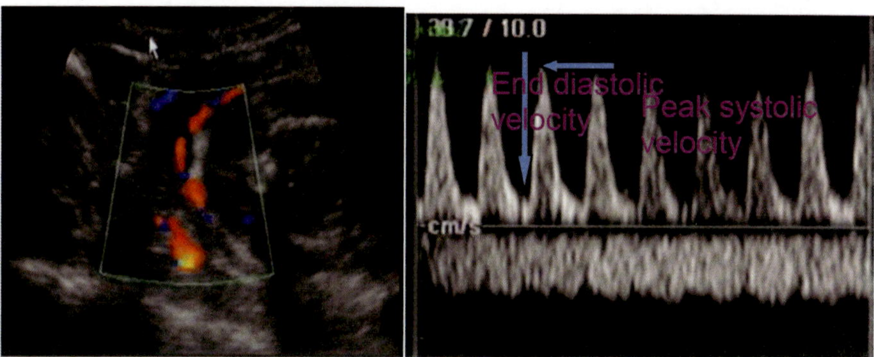

Fig. 7 & 8: is showing Doppler ultrasound on left and pattern of artery (above the baseline) and venous flow (below baseline).

Doppler ultrasound is, generally, used for haemo-dynamics. It has its usefulness in detecting pattern of blood supply to an organ. Generally, there are two colors; red for flow coming towards transducer and blue for blood flow going away from transducer. The blood supply to an artery comes in pulse form generally above the baseline and venous flow is uniform and shown below the baseline.

3

Setting of Ultrasound Machine for Optimal Imaging

Most of ultrasound machines are designed for imaging with a particular resolution linear and convex transducers. These machines can be used for general purpose for abdominal, obstetrics, gynecological, small parts, urology and cardiac applications. Most of the machines have standard measurement software viz. distance, area, volume, circumference, and obstetrical gestational age measurement for BPD, HC, AC, FL, GS, CRL, AD and FW.

Assessment of probes: One should examine the probe whether it is linear array or curved (convex) array. The probe should also be checked for its frequency. These days multi-frequency probes are supplied with the ultrasound-machine. These multiple frequencies could be 2.0 to 5.0 MHz with centrally fixed frequency of 3.5 MHz. The frequencies could be 4.0 to 7.0 (5.0 MHz setting), 5.0 to 9.0 MHz (7.5 MHz setting). The area and depth covered by various frequencies should be known.

Before setting the ultrasound machine, one should make sure of all the auxiliary devices for the machines. These are Thermal video printer, Ultrasound gel, Biopsy needle, image saving devices (USB device or internal saving mechanism), wire, Uninterrupted Power Supply etc.

Identifying and knowing the main system: One should note down the name and model of ultrasound machine as these are quoted in all the documents published with the ultrascan images generated on that particular machine. It is written on one side of front panel. On the front panel, there are various knobs for proper setting. One should locate the auxiliary power switch of the machine to begin with. Generally, there is power indicator on the machine that shows light on when power button is pushed to on position. The auxiliary power switch may also have indications of its ON and OFF positions, may be marked with 0 and 1 or other symbols. As shown in the figure below, there are various operating buttons on the operating panel. These might vary from Machine to Machine.

Operating Panel

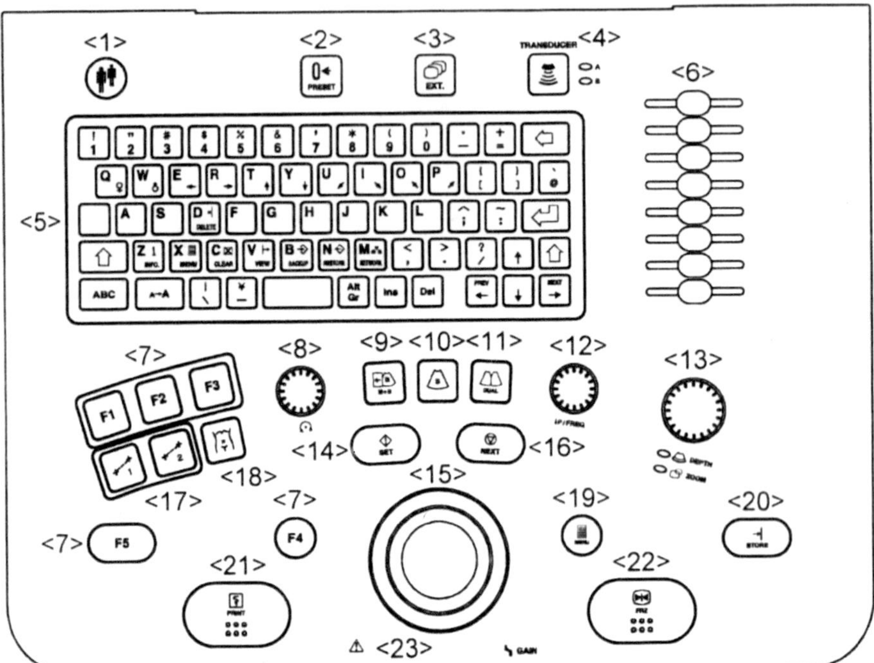

There are knobs for setting brightness and contrast of the machine. These may be on front side, bottom side or side wall of the machine. There are three additional settings for OVERALL-GAIN, NEAR-GAIN and FAR-GAIN. Moreover, on the side of display screen, there is bar having various divisions that are seen with proper setting. This bar is a standard to know proper setting. The machine should be set in such a way that all its division/segments are visible. In some machine this bar may be on top panel of the machine. All the ultrasound machines also have an attached keyboard. This keyboard is similar to computer keyboard with additional points on it. The additional point on the key board might be:

B: This button is used to set the machine in brightness mode with single screen (<10>)

B/B: This button is used to divide the display screen into two parts, active and frozen part. The active part is showing current scanning position, which can be frozen if required image is obtained. In this situation, other part will show active scanning. These two parts are switched between active and non-active sections (<11>).

B/M: This button is used to display brightness mode and motion mode. The motion section show moving lines, showing, if there is movement in the scan area, while brightness section show real time image of current scanning area (<9>).

M: This button is used to show motion on entire screen. These days B/M mode is commonly used than M mode.

All these buttons, when pressed can immediately switch to that particular mode. Freeze button: This is used to freeze the image and release the frozen image. There might be another button for freezing the image, which can be fixed to belt or pocket (<22>).

Direction: This button indicates the direction of transducer movement. Generally, there is point on the transducer that can be guide of direction of movement of image on screen, but this direction can be changed with pressing the direction button.

Print button: This button can be used to print the current image on the screen. Measurement button: There might be many buttons to do measurements on the scanned image. This includes buttons for distance, area, volume, trace, and other measure.

Track ball system/Mouse glide point: Every machine has a mouse, which can be ball or glide type. This ball or pad type point/area is used as mouse. In glide type, the pad can also be used for magnification.

Zoom in/Zoom out and Depth: There might be two functions in a button or separate buttons for separate function in some machines, which can be used for magnification in/out purpose.

Sockets: In addition to the above switches, there are many sockets to attach accessories to the ultrasound machine. These may vary from machine to machine, but many machines have most of them.

Foot-switch area: This can be used to attach additional control point from the foot.

Remote point: This can be used to attach remote switch for the control of machine.

Ground point: This point is used to avoid short circuit or shock to machine and operator.

Video in: This is used to attach wire to play an image on the machine. This is good to check the quality of recorded image on the machine.

Video out point: This point is used to attach printer or video cassette-recorder to record the ultrasound images. However, some machines have in-built floppy drive or videocassette system.

Focus buttons: There are four focus buttons: F1, F2, F3 and F4. These buttons can be used to focus the beam on a particular area. F1 focus the beam on uppermost part, while the F4 focus the beam on lowermost part of display. In some machine there might be sliding/rolling system to focus on a particular area. The focus arrow is shown on the side of the image either on right or left side. Sockets for transducers: In all the machines, there might be one/two/three/four sockets for attachment of probe(s)/transducer(s). There is a particular area/pocket on the machine, which is designed to place the probe, after use. Some machines also have a specifically designed place for keeping the ultrascan-gel.

System setup: The ultrasound machine should be kept at an optimal temperature ranging between 17 to 25 °C, with 40 to 60% humidity. One should avoid water vapor & splashing of water, direct sunlight, dust, high humidity, bad ventilation, salts and chemicals and gases. Before using the system for the first time, check that all the connections are securely plugged into their proper locations. The system should be properly grounded with the ground connector. The probe should be fixed in the probe connector and then turn the locking lever to lock the probe in place. Always turn the system off before changing or connecting the probe.

Cleaning and disinfection of probe: It is advised to use a sterilized probe sheath to avoid direct soiling of probe surface. These sheaths are disposable. Under field conditions good quality condoms or even disposable glove may be used, but insert some gel between probe and the layer of the cover. It is advised to thoroughly wipe the probe gently with warm-water, soaked soft cloth or use 3.4% glutaraldehyde solution for dipping the probe for 20 minutes for sterilization. Do not use other kind of cleaning system or solution. Do not drop or hit the probe against hard surface.

Attention during and after operation: Do not cover the machine or ventilating holes of machine during operation. If you turn the power off, wait for 5 seconds before turning the power on again. Do not touch the system or any accessories. Keyboard should be handled carefully. After operation, the power should be switched off using the button on the machine, before pulling the cord out of the main power connector. After cleaning the probe, store it in a specially designed probe case.

Identifiable points/information on ultrasound machine

Top panel or side panel of machine: Patient name, case number, active frame averaging, active probe type, frequency, Identification, Date & time.

Side panel or bottom: Surface depth, Scale marker, Active focal point marker, Current magnification factor, Active dynamic range, Active dynamic range.

Image control keys:

Print: Prints hard copies of screens

G.A.: Activates G.A. fetal weight, estimated delivery date calculation

I.D.: activates cursor to record patient name.

Menu: activates image processing, Setup, and switches system to external image source.

Measurement: Displays measurement menu according to active mode e.g. in B, B/B mode, distance; in M, B/M mode, velocity and hear rate calculation. CLEAR: Erase measurement results, body marks, and text information from the screen.

Exit: Exit a function or menu page.

Select: Specifies a function or item in a submenu.

Measure: Gets the results at each measurement and calculation

Indicator: Indicates the arrow mark at the specialized point during scanning.

Direction: When the image is live, this key changes the orientation of the screen image.

Special Key Functions of Keyboard Keys:

Document: Places system in documentation (text) mode.

Probe/Transducer: Changes active probe when using the external probe unit.

Body Mark: Displays probe placement and angle in the part of the body mark being scanned.

Focus keys: Selects active focal points.

Help: Display help message/

New-Patient: Activate new patient scanning.

Memory: Select current image memory

Report: Reporting measurements and calculations for a patient.

Frequency: Changing the diagnosis frequency.

Adjustment: Activate adjustment function for dynamic range and edge enhancement.

Trackball/Glide point: performs various functions depending upon image status and menu displayed e.g. when scanning live images, use the glide point to control depth in B mode and move M-mode-lines in B/M mode. When the image is frozen, it is used to position the cursor for measurements. This also controls the text cursor in Documentation mode and positions marker in body mark. To see upper portion of the image, roll the glide point up, while roll it down to see the deeper portion of the image.

Switching to an External Video Source: Press Menu go to setup; press select key and go to misc. key and then Ext. VDO. The screen will be ready to show display.

Storing the Image

Press freeze to freeze the image and then press save/memory key to store the image in memory of the machine. The image will be saved at Image Bank 1. Get another image and freeze and press save key to save another name. Interchange of the images is possible between two or more images. This allows comparing two pictures.

Image Retrieval: To see the stored image, press L (memory) to select the Image

Bank. The memory will be erased after power is off.

Image Processing

Gamma correction: Gamma correction provides Several choices to make the grey level of the image suitable for viewing on screen. Once setting of gamma level will be effective until reset again. Press Menu key, select imageproc, select gamma and then finally choose gamma 1, gamma2, gamma3 and gamma4. When picture quality is satisfactory, press exit to view on screen.

Frame averaging: This option is available in B and B/B mode and provides smoothness of image by specifying the speed of transition between the previous and current images. This reduces noises and speckle pattern on the screen. Four choices are available i.e. 1:0, 1:06, 1:1 and 1:1.6. The first number represents the previous image, while 2nd indicates current image. The higher the second number, smoother will be the image during scanning and movements of transducer. 1:0 should be used to view movements of internal organs viz.

Heartbeat. To use frame averaging: press Menu-key, select imageproc and select FRM.AVR. Each time the FRM.AVR is selected the portion of the bar in the screen will be filled. Press EXIT to come out.

NEGA: This function provides view of image in Black on white background. Press Menu key – select imageproc - - select NEGA. Doing again, changes the original position.

Dynamic Range: This adjusts the contrast of image. Press [A]-alphanumerical key to get ADJUST/DR/EE menu, select D.R. UP to increase the darkness, while select D.R. DN to make image brighter. There are 16 steps to use. Normally, machine is set with 4th step. EXIT key brings back to screen.

Edge Enhancement: This controls the edge contrast of the image. By controlling the edge enhancement, edge line of the image is clearly distinguished. There are 7 steps to change the edge enhancement. Press [A] alphanumerical key, select EE UP or EE DN to control the image sharpness.

Zoom: This can be used only in B/B mode only and provide 2 X zoom. Before using this function, select B/B mode. Press Menu-key, select imageproc, select zoom, and select the area on the image to be zoomed. Press select to zoom this area. Repeat the steps to zoom other area.

HISTOGRAM: This works with linear probe and shows echo pattern of selected area in graphic form. Press Menu key, select imageproc, select Histogram. The box and message "Area Move" appear on the image. Use track ball/pad to move the box to the area to get the histogram. Press SELECT to get display of "Area size". Use glide point/track ball to adjust the box size. Right movement: increases X-axis, Left movement: decreases X-axis, UP movements increase Y-axis, DOWN movement decreases Y-axis. Press SELECT to resize again. After fixing the size and location, press [MEASURE] to display information about histogram on right side of screen.

Total pixel number enclosed by the box

The dominant gray value in the box

The number of the dominant gray value enclosed by the box.

Different areas on the image can be selected go get various pixel values. WINDOW: This emphasizes or eliminates certain brightness value. There are 8 types of windows. Two of them are S-Curve and one R-S curve is fixed, while rest 5 are changeable and displayed in graph. Press Menu-key, select Imagproc, select Window. To use one of the curves, use glider point/track ball and Press SELECT. Text writing/documentation: To write desired information on the picture/screen, Press this key on box select "OFF or ON" to use free writing and or activate brief mode. Now position the cursor and start writing the text. Press "EXIT" or same key again to come out of the documentation. Brief- mode provides word within the memory of machine. It takes clue from the first two typed words e.g. typing of Ki shows kidney. Press docu, select "BRIEF" to activate the ON function. Press EXIT to return to scanning mode. In documentation mode, the key-board of machine acts like key-board of computer. To erase information from the screen, press backspace or CLEAR key. To remove any specific information about patient, press ID and then CLEAR/erase.

Body-Mark: This provides pictures of parts of body to be used for position and orientation of the probe. Several symbols for internal medicine and many for Gynecology & Obstetrics and also several symbols for specialized fields might be available. Press body mark key, select the desired body parts. A box with options appears. Use the track ball/glide point to position the probe symbol corresponding to the actual scanning location. Use Glide pointer/trackball to position the probe symbol at desired location. Press [EXIT] or Body mark key to come out of the function. Press the [CLEAR]/ERASE key to erase the body mark. EDITING REPORT: The machine provides information about G.A. and one can enter SEX, AGE, DR, Names for all reports. Report can be erased.

Note: The keys and functions vary with machine to machine.

4

Anatomy of Some of the Organs in Dog and Horse in Relation to Ultrasonography

In order to evaluate the correct status of the organs their correct disposition within the body becomes very essential. Although one cannot see the organs beyond the skin, but appropriate knowledge of the topography of the organs can act as a glass covered watch wherein one can see from outside the arms and the dial of the watch.

Horse

Heart: The heart is placed in the middle mediastinum of the thoracic cavity with its long axis directed caudo-ventrally. Three fifth of the heart lies on the left of median plane while two fifth is on the right of the median plane. The base is placed dorsally, with its highest point at the junction of the dorsal and middle thirds of dorso-ventral diameter of the thorax. As a whole it lies opposite to second intercostal space/third rib to sixth rib/sixth intercostal space. The apex of the heart lies in the center of the last segment of the sternum about half inch from the sternal part of the diaphragm. Heart is directly related to the lateral thoracic wall at the cardiac notches formed in the lungs. The left surface of the heart extends from third to sixth rib in the lower third of chest while the right surface extends from third to fourth intercostal space.

Pulmonary Valve: It can be better visualized from the left side in the third intercostal space: a hand breadth ventral to the horizontal line passing through the shoulder joint.

Aortic Valve: The position of the aortic valve is on/or ventral to the horizontal line passing through the shoulder joint on the left side in the fourth intercostal space.

Left Atrio-Ventricular Valve: It is placed hand breadth ventral to the horizontal line passing through the shoulder joint in the fifth intercostal space on the left side OR 12-15 CMS dorsal to the olecranon process.

Right Atrio-Ventricular Valve: It can be better visualized from the right side in the fourth intercostal space hand breadth dorsal to the olecranon process while from the left side, hand breadth dorsal to the olecranon process in the third intercostal space.

Kidneys & Urinary Bladder: Kidneys are retroperitoneal and are placed below the vertebral ends of the ribs. The left kidney is bean shaped and its length is 15-20 cms while the breadth is 11-15 cms. The caudal pole is wider as compared the cranial pole and the ventral surface is more convex than the dorsal surface. It extends from seventeenth (16-18) rib to the second or third lumbar vertebra. The right kidney is heart shaped with its base directed medially. The breadth is 15-18cms while the length is 13-15cms. The cranial pole is convex while the caudal pole is almost straight. It extends from sixteenth rib to first lumbar vertebra.

The urinary bladder is placed on the floor of the pelvic cavity and its position varies with its fullness. When completely empty it is placed entirely in the pelvic cavity and when it completely distended it may extend up to the umbilicus.

Spleen: Spleen of horse is roughly triangular (Scythe) in shape with its base dorsally placed and the apex directed ventrally. It is very closely related to the greater curvature of the stomach and left crus of the diaphragm. The cranial border is thin and concave while the caudal border is convex. The dorsal extremity or the base is related to saccus caecus and left kidney and extends from seventeenth thoracic vertebra to upper part of the flank just behind the last rib or just caudal to the transverse process of the first lumbar vertebra. The caudal border is initially parallel to the costal arch, but at the level of sixteenth intercostal space it follows a line connecting the olecranon process to the tuber coxae. The ventral extremity is placed opposite to the ninth to eleventh rib about a hand breadth above the costal arch.

Stomach: The stomach of horse is roughly 'U' shaped with its right limb slightly shorter than the left limb. The major part of the stomach lies to the left of the median plane; however, the pyloric part is placed right to the median plane. Cardia in a moderately filled stomach is present about an inch or more to the left of median plane and ventral to the vertebral end of the fourteenth rib. Saccus caecus, the left rounded extremity of the stomach lies ventral to the left crus of the diaphragm below the sixteenth or seventeenth rib's proximal part. The pyloric extremity is placed just right of the median plane about 2 inches lower than the cardia.

Liver: Liver is placed mainly on the right side of the body. In older horses, the right lobe is atrophied as a result of pressure of right dorsal colon in which case the left lobe is larger than the right lobe. Sometimes the left lobe is atrophied due to pressure of stomach and in that case the right lobe is larger than the left lobe. Three fifth of the liver lies on the right side while two fifth is present on the left side. The long axis of the liver is obliquely directed from right to left in a cranio- ventral direction. The ventral border of the liver is placed dorsal to the sternum about three to four inches from the floor at the level of sixth to seventh ribs. The right border ascends obliquely in a caudo-dorsal direction to a point about ten cms dorsal to fifteenth costochondral junction. From here it ascends dorsally following the fifteenth intercostal space and widens to form the renal impression before reaching the vertebral column. The left border arises from the ventral end of sixth to seventh ribs about three to four inches from the floor and passes dorsally and slightly caudally to the dorsal end of tenth intercostal space. The dorsal border is thick and along this border runs the caudal vena cava.

Dog

Heart: The heart in dog is placed very obliquely. The base is present opposite to the ventral part of the third rib. The apex lies in the sixth inter-chondral space or seventh cartilage towards the left side and is in close contact with sternal part of the diaphragm. The cardiac notch where the heart is closest to the lateral thoracic wall on the left is between fifth and sixth inter-chondral space while on the right it is between fourth and fifth inter-chondral space.

Kidneys & Urinary Bladder: The kidneys of dog are bean shaped. Right kidney is placed opposite to the bodies of first to third lumbar vertebrae and may extend up to the last thoracic vertebra. Cranial half of the kidney lies in the renal fossa of the liver. The left kidney is loosely attached to the peritoneum and its position varies with the fullness of stomach. In the empty stomach animal it lies between second to fourth lumbar vertebrae.

The urinary bladder lies on the ventral abdominal wall cranial to the pubis, however, the neck is intra pelvic in position.

Spleen: Position of the spleen to a great extent depends on the fullness of stomach. In empty stomach the spleen is placed within the costal arch. In a moderately filled stomach, the dorsal end is covered by the last two ribs and the ventral-end projects beyond the costal arch and reaches ventrally as far as the levels of the ventral end of seventh to tenth ribs and caudally to transverse plane through the second to fourth lumbar vertebrae. In a greatly distended stomach the spleen is present in the flank region and may at times reach the pelvic inlet.

Stomach: The empty stomach cannot be palpated because it lies in the intra thoracic part of the abdominal cavity. It fails to make contact with the ventral abdominal wall and is placed at the level of tenth costochondral junction. Fundus is in contact with the diaphragm below the angles of tenth, eleventh and twelfth ribs and the cardia is placed left of the median plane ventral to eleventh thoracic vertebra. . The pylorus lies against the hepatic portion the right, opposite to middle of tenth rib. In a moderately distended stomach it enlarges cranially and caudally from ninth rib to first or second lumbar vertebrae. Ventral parts of the body and parts of the pyloric antrum leave the intrathoracic part of the abdominal cavity and make contact with the floor of the abdominal cavity. The greatly distended stomach may extend up to the pelvic inlet and is contact with the ventral abdominal wall.

Liver: The liver of the dog is lying almost on the median plane except that when the stomach is full it is pushed slightly towards the right of the median plane. The ventral border is lying just caudal to the xiphoid cartilage. The left border is at the level of tenth intercostal space or eleventh rib. The right border is placed along the costal arch and the end of the caudate process is ventral to the right kidney; opposite to or little behind the last rib.

5

Comparison of Radiography and Ultrasonography in Diagnosis of Thoraco-Abdominal Cases

The affections of abdominal cavity incidence wise are on the top followed by thoracic cavity. Both these affections lead to high morbidity and heavy mortality losses. Most thoracic affections produce almost similar clinical signs but abdominal affections can show different clinical signs. Radiography and ultrasonography play an important role in differentiation and confirmation of various affections.

Radiography is an important diagnostic tool for both thoracic and abdominal diseases in small animals but for large animals it is more useful for thoracic affections as compared to abdominal affections. In large animals, the use of radiography is limited up to the ventral aspect of the abdomen only i.e. for diagnosis of diseases of reticulum, diaphragm and sternebrae. In thoracic cavity, there are numerous diseases that need to be differentiated. Pneumonia, lung cyst, lung abscesses, pleuritis and pleural effusions are some important diseases apart from presence of foreign bodies. In such a situation, radiography can play an important role as an aid in the diagnosis of various affections.

Similarly ultrasonography has also played a major role in the diagnosis of abdominal diseases in small animals such as dogs and newborn calves especially in the field of G I tract and urinary system. The ultrasonography is advantageous over radiography in the sense that internal anatomy of the soft organs can be better known and studied by ultrasonography as compared to radiography. It is also considered to be extremely safe for both the patient and the operator.

Most organs are well suited to ultrasound images except bones and gas containing structures such as lungs, stomach and bowel. Bone and gas reflect ultrasound and do not allow the sound waves to image structures distal to them therefore radiography gives better diagnostic images for these organs. The liver, spleen, kidneys, prostate, gravid uterus and fluid containing organs like urinary bladder, gall bladder and major blood vessels can be readily imaged with ultrasound. Keeping in view all these points a comparison of both techniques has been made to know which technique is more suited for which structures/organs.

Different body organs/ structures can be divided into five main components such as air, water, fat, cartilage, bone and of course metallic objects if present on the basis of their densities. The radiographic and ultrasonographic images of these components is different as shown below:

1. **Air/ gas:** Lung and G. I. tract fall under this category. These organs have negative density & Absorb least amount of radiation and causes ionization of silver and appear black on radiograph but no formation of echo by ultrasound waves hence image will be anechoic.
2. **Water:** Muscles, blood and various organs having water comes under this category. These structures absorb less amount of radiation and will appear darker gray on radiograph. These organs are best suited for ultrasonography and echo produced will be hypo and will appear black on ultrasonogram.
3. **Fat:** The absorption of radiation is still less therefore image will be gray on radiograph and will be iso-echoic on ultrasonogram.
4. **Cartilage:** Image on radiograph will be gray just like fat but echo produced by ultrasound wave will be hyper and will give white image.
5. **Bone:** Radiation absorption is more and forms white image on radiograph. The ultrasound wave gets reflected hence echo production will not be proper and the image will be white with hazy boundary.
6. **Metal:** Complete absorption of radiation and will prevent ionization of silver therefore image will be pure white on radiograph. More reflection of ultrasound waves and image will be hazier on ultrasonogram.

Pathological lesions also can be classified depending on their densities e.g. radiopaque or radiolucent and hypo or hyperechoic and reflect images as per their densities e.g. abscesses, cysts and tumor reflect the shadow of air and water or both. For diagnosis and comparison purpose the body can be divided into two main components:

1. Thoracic cavity.

2. Abdominal cavity

1. Thoracic cavity

The thoracic cavity can be seen on radiograph bounded dorsally by the thoracic vertebrae, laterally by ribs, ventrally by sternum and sternal ribs. cranially by thoracic inlet and caudally by diaphragm. Radiographically, only a small part of the diaphragm is visible. Its appearance is variable depending upon the phase of respiration, age, degree of intra-abdominal pressure and direction of the primary X-ray beam. On lateral radiographs, the diaphragm will appear as convex structure with its convexity toward thorax.

The normal lungs are composed primarily of air. The remaining portion of the lung is the interstitium, which contains bronchi, bronchiales, alveoli, lymphatics, nerve and muscle. Basically, two radiographic densities are visible in the lung field, i.e. air and soft tissue. The pleurae are the membranes that cover the lungs and line the thoracic cavity. The pleurae cannot be seen on lateral radiograph but the space between the two-pleural sacs can be visualised because of the structures within it.

Distribution of thoracic disorders for the last 20 years in relation to involvement of anatomical structures revealed that lung lesion occupies a unique position (70%), followed by heart (25%). The pathological lesions of the thoracic wall and sternum accounted for low occurrence (5%).

Disease wise distribution showed maximum cases of pneumonia (48%) followed by pericarditis (20%), lung cyst (10%), lung abscesses (5%) and cavitary lesions (4%). Other affections include pleuritis, foreign body lodgement, pleural effusion, diffuse fibrosing alveolitis, pneumothorax and diaphragmatic abscess.

For thoracic ultrasonography, heart is the common structure, which is imaged. It is best imaged by echocardiography especially in horses. Echocardiographic exam is important in horses because physiologic flow murmurs are commonly auscultated & may be difficult to distinguish from murmurs of valvular regurgitation. Abnormal cardiac functions can be detected even without significant murmurs. Ultrasonography should be part of complete work-up in animals presented for poor performance because normal cardiac function is crucial for superior athletic performance

Scanning Technique

Entire heart should be imaged from one cardiac window so that cardiographer can best appreciate relative sizes of cardiac chambers & vessels and their relationship to one another and to cardiac valves. A sector transducer is ideal because the small transducer footprint is maneuverable in small inter-costal space of horse & it has up to 30cm penetration.

2. Abdominal cavity

The abdominal cavity is very voluminous in bovine and it is not possible to radiograph such a thick area only cranioventral portion of the abdomen is proper in thickness to be radiographed. Therefore main area of study is always cranioventral abdomen. On lateral radiograph of the cranioventral abdomen, the parts discernible are caudal part of the sternum, ventral portion of the diaphragm, reticulum and occasionally part of the abomasum. Caudal border of the heart and ventral part of diaphragmatic lobe of the lung can also be seen. The diaphragm will appear as smooth line separating the abdominal and thoracic cavities joining the sternum just cranial to the 7th sternebra during the peak inspiratory phase.

A normal reticulum will appear as a soft tissue density lying adjacent to the diaphragm and along the ventral abdominal floor in the sternal region. It lies between the 6th and 9th ribs. On most of the time the interphase between the reticulum and the diaphragm is difficult to appreciate. The ventral boundary of the reticulum lies just dorsal to the sternum or in contact with the ventral abdominal floor, the tip just caudal to the xhipisternum. This position of the reticulum is variable depending upon its contractory or relaxed phase. Position of the reticulum can easily be marked if some sorts of debris or foreign bodies (metallic or non-metallic) are present.

Distribution of abdomen disorders for the last 20 years revealed that reticular lesion occupies a unique position (50%), followed by diaphragm (35%). The pathological lesions of the sternum accounted for low occurrence (.5%). The remaining cases observed are lodgement of foreign body at different locations of the abdomen, septic peritonitis, formation of fistulae\tracts at different site of the abdomen etc.

Disease wise distribution showed maximum cases of foreign bodies (45%) followed by diaphragmatic hernia (30%), extra-reticular abscess (10%), septic/ aseptic peritonitis (5%) and reticular adhesions (4%). Other affections include fistulous sternebrae, reticular fistulae etc.

Recently various workers have tried ultrasonography for various diseases of reticulum such as reticular abscess in cows and buffaloes. It has been observed that the reticulum can be depicted as homogenous oval structure toward ventrolateral side of the abdomen. In cases where there was formation of an abscess between reticulum and abdominal wall and there was adhesion formation, the ultrasonogram showed a hypoechogenic cavity surrounding by echogenic capsule. The contents of the abscess appear heterogenous and are partitioned by echogenic septa.

In small animals, all abdominal organs can be imaged by radiography but only their external shape and size can be judged but ultrasonography is very useful for studying internal details of these organs especially of organs like liver, kidney, stomach and intestine containing fluid, urinary bladder and urethra.

1. Scanning of the Liver

For scanning of the liver, the area between the xiphisternum and the umbilicus including the area on both sides of the ventral midline was shaved properly. The liver was scanned with the animal controlled in dorsal recumbency. The transducer plane was aligned perpendicular to the bodyline just behind the xiphisternum over the midline in the craniodorsal direction to get a transverse scan of the liver. The ultrasonic images were freezed at maximum expiration to maximize the distance between the diaphragm and the skin and in this position:

1. Circumferential diameter of gall-bladder was measured.
2. A single linear hepatic measurement from the ventral lobes of the liver up to the inner limit of the diaphragmatic line through the gall bladder was taken. Measurements were correlated with age of the animal.

2. Scanning of the Kidneys

For scanning of the kidneys, the area between the lumbar vertebrae and the ventral midline was shaved on both sides of the body including the area over the last two intercostal spaces. The animal was controlled in lateral recumbency i.e. right lateral recumbency for scanning left kidney and left lateral recumbency for scanning of the right kidney. Left kidney was scanned below the transverse processes of first to the third lumbar vertebrae and for the right kidney, an intercostal approach through the last or second last intercostal space was used to scan it in its transverse plane. Each kidney was scanned in its transverse plane, a longitudinal and a sagittal plane Appearance of a C-shaped image in the middle of the kidney indicated a proper transverse plane. In the longitudinal plane, a bright renal sinus was visible in the centre or eccentric location of the kidney. In the sagittal plane, two bright parallel echogenic bars were visualized in the centre of the kidney. The following measurements of both the kidneys were taken:

1. **Kidney length (mm):** The distance between the cranial and caudal ends of the kidney was considered as the kidney length. The kidney length was measured in the longitudinal scan.
2. **Kidney height (mm):** The kidney height was measured from the dorsal to the ventral border of the kidney at two points just cranial and caudal

to the renal sinus. The mean of these two measurements was considered as the height of the kidney. The kidney height was also measured in the longitudinal scan.

3. Scanning of the Stomach

The stomach of the dog lies under left hypochondriac region. Usually the empty stomach can not be palpated. The moderately filled stomach lies between 9 -12 rib to 1-2 lumbar vertebrae. The full stomach present up to 3 to 4 lumbar vertebrae.

The animal can be put in dorsal, left or right recumbency or in standing positon. Different section-transverse, longitudinal & obliques can be taken using real time sector scanner- 3.5 to 7.5 mz. Normal ultrasonographic appearance and cross-sectional images of different portion show that ultrasound wall contain five layers such as mucosal surface, mucosa, submucosa, subserosa and serosa. Normal luminal patterns

Mucous – echogenic contents without acoustic shadowing

Fluid – anechoic luminal contents & act as window.

4. Imaging of Urinary Bladder & Urethra

The urinary bladder and urethra can be visualized both by radiography and ultrasonography in small animals especially in dogs. The conditions of urinary bladder such as stone are best imaged by radiography. The cases of cystitis and other inflammatory conditions can be better seen by sonography. The changes in density of urine can also be better judged by ulrasonography. For imaging of urinary system both techniques are useful.

6

Proper Transrectal and Transabdominal Ultrasound Procedure in Cattle and Buffaloes

General Instructions

The monitor of ultrasound machine should be placed at the eye-level. The daylight should not be too bright. The setting of the machine should be done to obtain best picture on the screen. There are various knobs on the machine to set contrast and brightness. There are three other knobs that are used to set near gain, far gain and overall gain. There is also a brightness bar (help-bar) on the screen with different brightness areas. These knobs should be turned clockwise or anti-clockwise, so that all the areas on the help bar (brightness bar) are visible.

Transrectal Procedure

In cattle, trans-rectal ultrasonography is done in a way, similar to rectal examination done for palpation of genital organs. Following steps are suggested for proper procedure.

1. Properly restrain the animal.
2. Evacuate the feces from the rectum.
3. Select a transducer of proper frequency (the linear array transducer of 5.0 MHz frequency is useful for general purpose in bovines, however for detailed examination of early conceptus and small follicles, a transducer of 7.5 MHz can also be used. Some transducers have range of these frequencies in single-equipment. Such transducers of switch-able frequency are found more useful.
4. Apply ultrasound gel on the transducer, before insertion into rectum. Some manufacturers advise to apply a cover over the transducer to avoid direct soiling of active surface of the transducer. The covers, provided by the company, are of good quality. If proper cover is not available, then a disposable glove can also be used. If scan is for assessment of

intensity of the organ/tissue, then avoid using these covers, particularly disposable gloves.

5. The gel-lubricated transducer should be held in a hand folded in a cup shape and is taken through anal opening. It is advised that the transducer should be held in such a manner that all structures to be examined are first felt by the hand. Some practicener prefer to conduct palpation of the reproductive organs, before insertion of transducer into rectum.

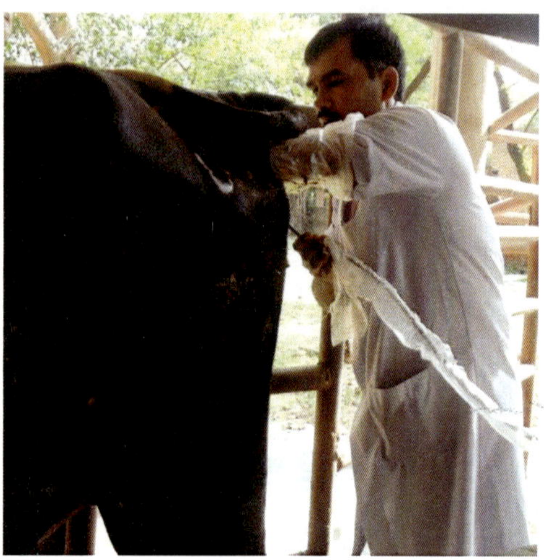

Photograph. 4: is showing transrectal ultrasonography in a cow

6. The active surface of transducer should be pressed firmly against the ventral mucosa to ensure the transmission of ultrasound waves through rectal mucosa into abdominal viscera. The urinary bladder is taken as landmark for scanning of reproductive tract. The urinary bladder will appear homogeneously non-echoic i.e. black. The images of uterus are obtained just cranial to the urinary bladder. The uterine body, cervix and the vagina are imaged in long axis view (as these are midline structures), as the transducer is moved along the rectal floor.
7. The images of uterine horns and ovaries are obtained, while moving
8. Transducer in lateral and slightly downward direction by changing the angle of the transducer. It is suggested that the reproductive organs should be scanned in a sequence viz. vagina, cervix, body of uterus, uterine horns and the ovaries.
9. The images of ovaries can be easily identified, if there are follicles on its surface. For a beginner, there might be some problems in identifying the

images; therefore, it is advised to practice to identify these images on a genital organ from slaughterhouse in a water bath.

10. The transducer should be moved slowly. The good images should be frozen with the help of a button on the machine or a remote button for taking the print out or for recording on a VCR.

11. For images of fetus in pregnant uterus, the hand of ultrasonographer having probe in hand should be taken towards ventral side of horn. From ventral side image in cow appears better than dorsal surface.

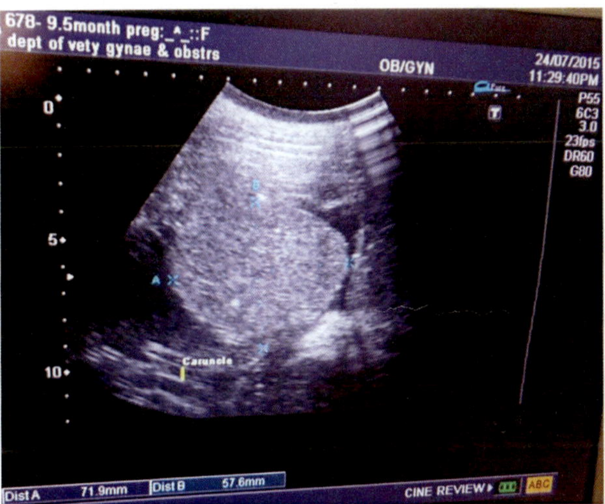

Fig. 9: is showing cross sectional full view of placentome in a buffalo. It measured 71.9 mm x 57.6 mm at 9.5 months of pregnancy.

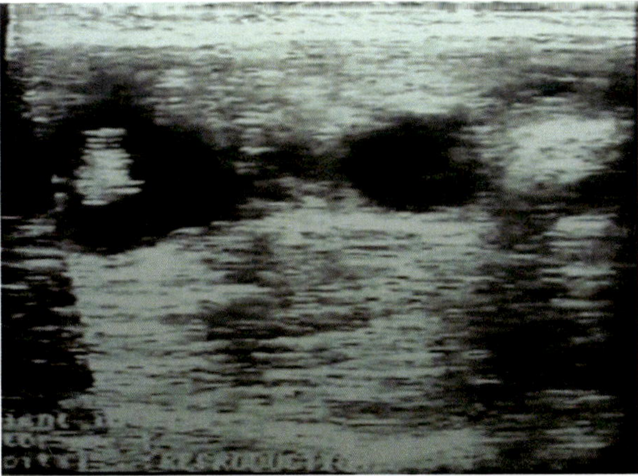

Fig. 10: is showing pregnant horn with fetus on day 29 in a cow imaged with 7.5 MHz transducer, transrectally.

Procedure of Transabdominal Ultrasonography

There are difficulties in assessing advance pregnancy with rectal palpation particularly when uterus is located at abdominal cavity. In the cases suffering from uterine torsion, it is not possible to palpate the foetus. Transabdominal ultrasonography is very useful to diagnose the pregnancy at the period when the fetus is not approachable by per-rectal method. Additionally, it can help to know viability of fetus, to image foetal organs and to rule out any pathological abnormalities of pregnancy. In the buffaloes, trans-abdominally transabdominal ultrasonography can be done by using 3.0 MHz frequency with a convex transducer. The transducer can be applied near udder in the ventral and lateral abdominal region.

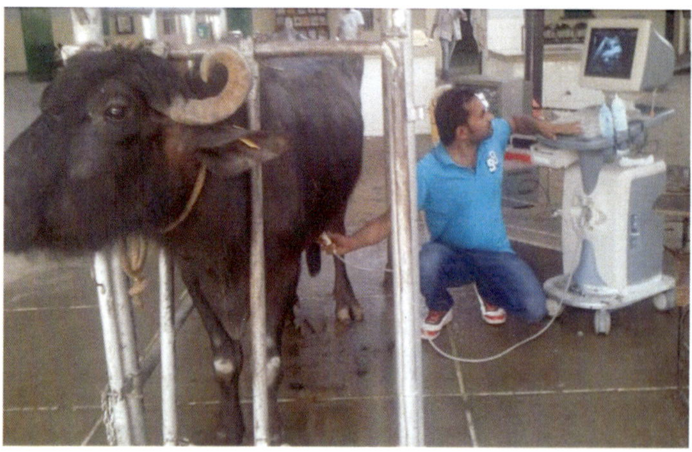

Photograph 5: is showing transabdominal ultrasonography in a buffalo

In authors' experience, the status of fetus could be easily detected by imaging the foetal heartbeats in advance pregnancy. In some cases, it required to change side of scanning, as images of foetus were not always possible from same side. Proper shaving of site is helpful in getting good ultrasonic images. From transabdominal ultrasonography fetal body parts like limbs, head and neck; fetal skeleton like ribs and vertebral column, the visceral organs of fetus like liver, heart, kidney, stomach and umbilicus are easily visible.

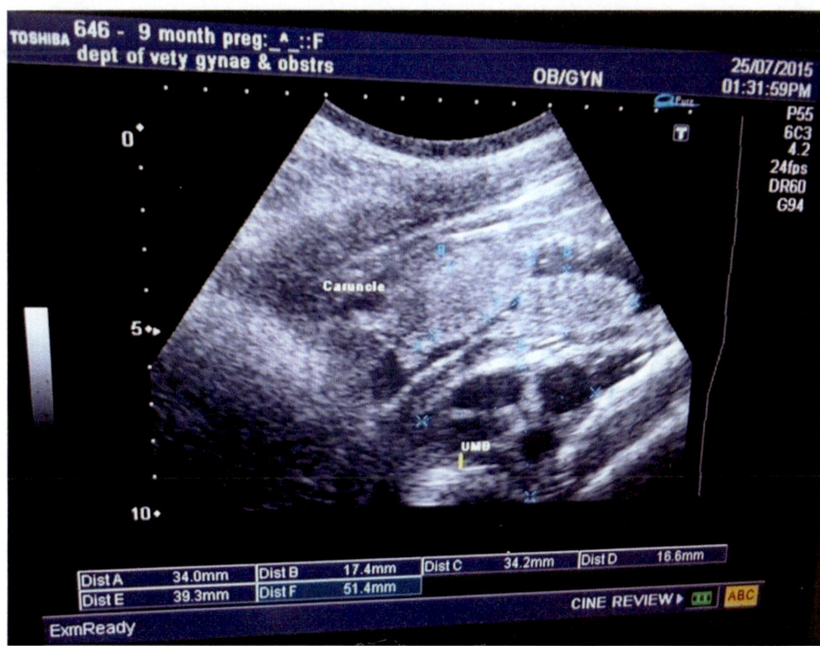

Fig. 11: Showing dimeters of umbilicus in a buffalo. There are four vessels in the umbilicus. Two arteries and two veins. The arteries are larger in diameter and veins.

However, one may face difficulties in imaging internal organs of fetus in uterine torsion cases. With experience, the trans-abdominal ultrasonography could be very useful tool for veterinarians for differential diagnosis of normal and abnormal advance pregnancy in cows and buffaloes.The transabdominal ultrasonography can be used in buffaloes from 5 months of gestation, while in cross bred cows, it can be used after 4 months of gestation. Around that time the position of probe will be lower side of paralumber fossa. The position of probe will go down with advancement of pregnancy.

7

Reproductive Ultrasonography in Buffalo

Failure to accurately diagnose ovarian function is said to be one of the reason for poor reproductive efficiency in this species. The information regarding ovarian function can be gathered either by observing the reproductive behavior of the animal, status of reproductive hormones or palpation of ovaries per rectum for the presence of cyclic or pathological structures. However, due to poor expression of estrus signs behavioral observations poorly reflect the accurate information about ovarian function in buffaloes. Similarly, palpation per rectum does not provide reliable diagnosis since the CL is frequently embedded deep in the ovarian stroma of buffalo ovary and findings vary due to its subjective nature. Furthermore, measurement of hormones requires sophisticated laboratory procedures/ instruments for making a correct diagnosis. Therefore, real time ultrasonic examination is a better alternative to ascertain reproductive status of buffaloes and for devising suitable strategies to optimize fertility of this species (Sharma, 2003). However, it is important to mention that information gained from ultrasound must be used in conjunction with reproductive history of the animal, palpation of genital tract, and other diagnostic procedures rather than in isolation.

Ultrasonic Appearance of Ovarian Structures

Ovarian Stroma: Ovarian stroma is more echogenic compared to most other ovarian structures (follicles / CL) and therefore, appear brighter on the viewing screen. Occasionally, however, ovarian stroma is difficult to appreciate, due to the fact that luteal tissue occupies almost the entire ovary, especially in buffaloes. Identification of ovarian stroma is aided by the presence of numerous scattered non-echogenic follicles, which appear as round and black areas surrounded by thin wall. Follicles as small as 2 to 3 mm diameter can be visualized, counted and measured by ultrasonography. Maximum diameter of the pre-ovulatory dominant follicle (DF) at oestrus ranges from 10-14 mm in buffaloes.

Ovulation: Ovulation can be detected by sudden disappearance of the large pre- ovulatory size follicle on subsequent ultrasound examination. It is also possible to detect the corpus haemorrhagicum on the first day after ovulation. The developing corpus haemorrhagicum is a poorly delineated, irregular, slightly echogenic grayish black structure within the ovary.

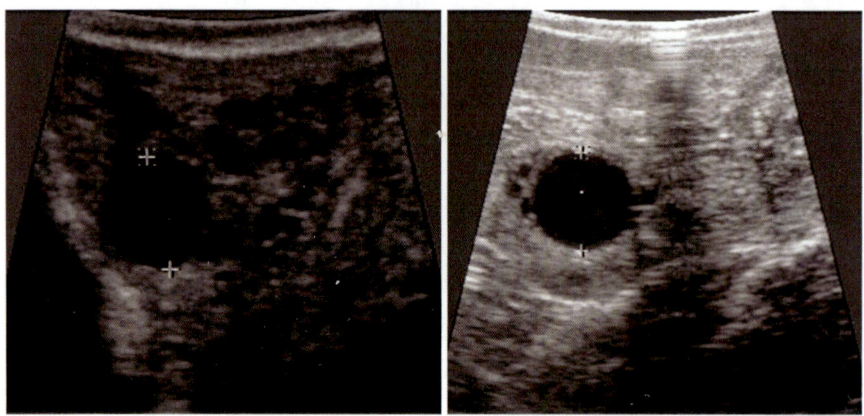

Fig. 12: is showing dominant follicle of 14.2 mm and other small follicles (left) and 16.5 mm ovulatory follicle (right) in a buffalo

Corpus luteum: A developing CL is an irregular mildly echogenic structure, while mature CL is more echogenic and appears as a well defined granular oval structure with a line of demarcation from ovarian stroma. In the regressing CL, the line of demarcation becomes faint due to minor differences in echogenecity from that of the ovarian stroma. In buffaloes, the size of CL ranges from 18-25 mm in diameter.

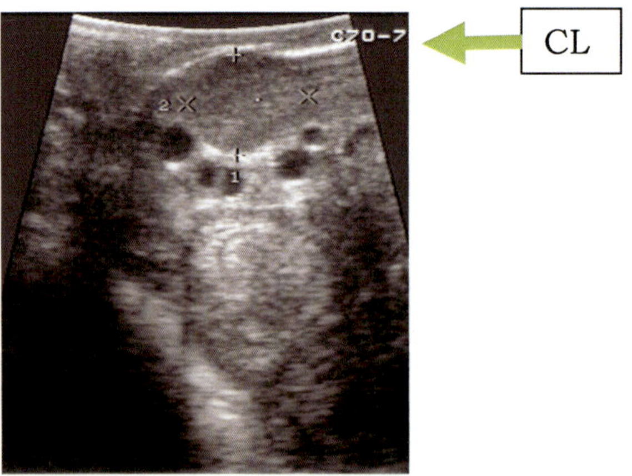

Fig. 13: is showing Corpus luteum of 21.5 mm along with small follicles (black) on the periphery of an ovary of buffalo

Ovarian cysts: Ultrasonography of ovarian cysts presents large (25-55 mm), non- echogenic round structures (black) that are either single or multiple large follicles of variable shape and size.

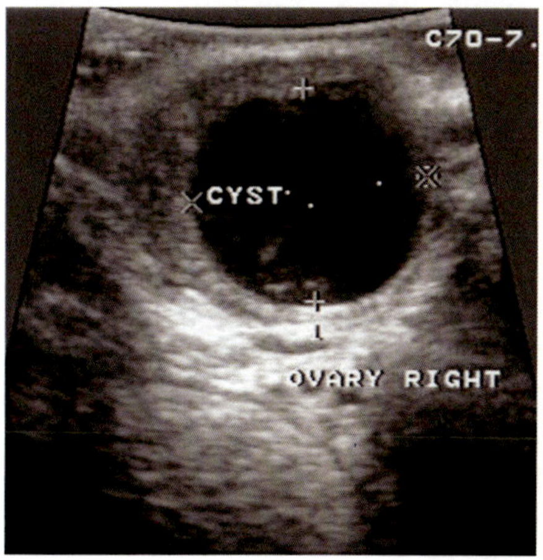

Fig. 14: is showing follicular cyst (size 22 mm X 25 mm, approx.) on ovary of a buffalo

Differential diagnosis between a follicular cyst and a luteal cyst can be made on the basis of thickness of the cyst wall. The thickness of the wall of a follicular cyst is <3mm while it is >3mm in case of luteal cyst. A small fluid filled cavity may be present in the corpus luteum. Such condition is called cystic corpora lutea and is non pathological. Parovarian cysts appear similar to the follicular cysts but are located outside the boundary of ovarian stroma.

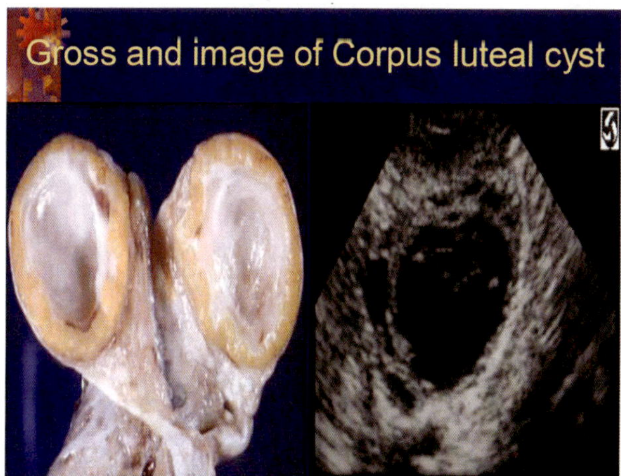

Photograph 6: and figure 15 are showing photograph of cyst (left) and ultrasonic image of luteal cyst (right).

Follicular Dynamics

Sequential ultrasonic monitoring of individually identifiable follicles is a powerful tool for studying follicular dynamics during the estrous cycle, pregnancy, postpartum period, superovulation and in relation to hormonal treatments. Ultrasonography can also be used to monitor the growth and regression of ovarian follicles. It is now well established that follicular growth occurs in a wave- like pattern. A follicular wave involves synchronous growth of a group of follicles in both ovaries, from which one follicle attains dominance over others to become the dominant follicle (DF). Each DF has a growing, static and regressing phase. There may be one, two or three waves of follicular growth during estrous cycle in buffaloes (Baruselli *et al.*,1997). The first wave begins on day 1, the second around day 9-11 while the third wave appears on day 17 of the estrous cycle. Only the DF of the final wave ovulates, whereas the DFs of the preceding waves undergo atresia. Sequential ultrasonic monitoring of individually identifiable follicles is a powerful tool for studying follicular dynamics during the estrous cycle, pregnancy, postpartum period, superovulation and in relation to hormonal treatments.

Uterus and Cervix

Pathological and other physiological conditions of uterus can be identified using ultrasonography in large animals (Fissore *et al.*, 1986). Shape and echotexture of the uterus varies during the different phases of estrous cycle. It is possible to distinguish endometrial folds and myometrium as well as small accumulations of fluid in uterine lumen, during estrus. However, during diestrus, fluid is absent and uterine wall is less distinctive in shape and size. Endometrium can be distinguished from more echogenic myometrium. Ultrasonic image of cervix on day 0 is characterized by presence of non-echogenic fluid collections within a hyperechogenic image. However, during diestrus phase non echogenic area is seen. Uterine pathological conditions, such as endometritis, pyometra, mucometra and mummified/ macerated fetuses are generally characterized by a thickened uterine wall and a distended lumen, filled to varying degrees with partially echogenic snowy patches. In fetal maceration, the fetal bones are identifiable as echogenic particles in the uterine lumen suspended in the fetal fluids. In mummified fetus, the uterine fluid is absent and fetal mummy appears as a poorly defined echogenic mass.

Early Pregnancy Diagnosis

The greatest advantage of ultrasonography is diagnosis of early pregnancy in buffaloes. With this technique confirmed diagnosis of pregnancy can be made as early as day 25 post-insemination. However, the accuracy widely varies depending on a number of factors, including the stage at which pregnancy is examined, experience of the operator, frequency of transducer selected, age and parity of animal and number of times animal is examined. Early diagnosis of pregnancy is based on the detection of a discrete, somewhat linear non-echogenic structure within the uterine lumen. However, the diagnosis based solely on detection of fluid within the uterine lumen is unreliable on two counts: 1) the small dimensions of the elongated conceptus approach the limits of resolution of most available scanners and 2) small amount of apparently normal free intrauterine fluid are present as early as D 10 and can be indistinguishable from the vesicle. The embryonic vesicle can be observed between day 18-20 and fetal heart beat is detectable between days 21-22. Our personal experience suggests that pregnancy can be diagnosed in buffaloes with certainty on day 25 post breeding. Furthermore, the stage of gestation can be ascertained by measuring a variety of parameters like crown-rump length, transabdominal diameter and biparietal diameter (Kahn, 1990).

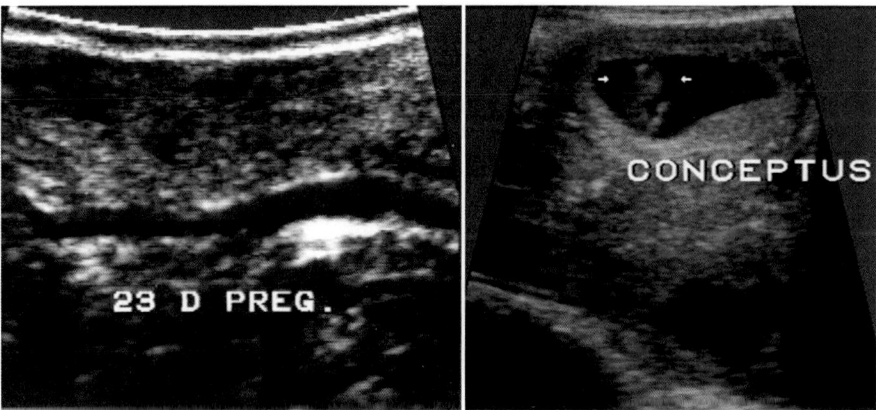

Fig. 16 & 17: are showing accumulation of fluid (left) on day 23 and conceptus on day 35 (right) of pregnancy in a buffalo

Fetal Sex Determination

One of the important applications of ultrasonic imaging in fetal studies is the development of a highly accurate technique for the ultrasonic diagnosis of fetal sex in equines and bovines. Fetal sex determination has the economic advantage for animal owner since a buffalo carrying female fetus is likely to fetch more prices compared to that carrying a male fetus. Similarly, it also helps in deciding whether or not to retain a pregnant buffalo already earmarked for culling, besides being useful in progeny testing program. In bovines, diagnosis of fetal sex is based on the presence of scrotal swelling and mammary glands or the location of genital tubercle which leads to the formation of penis or clitoris (Curan, 1992).

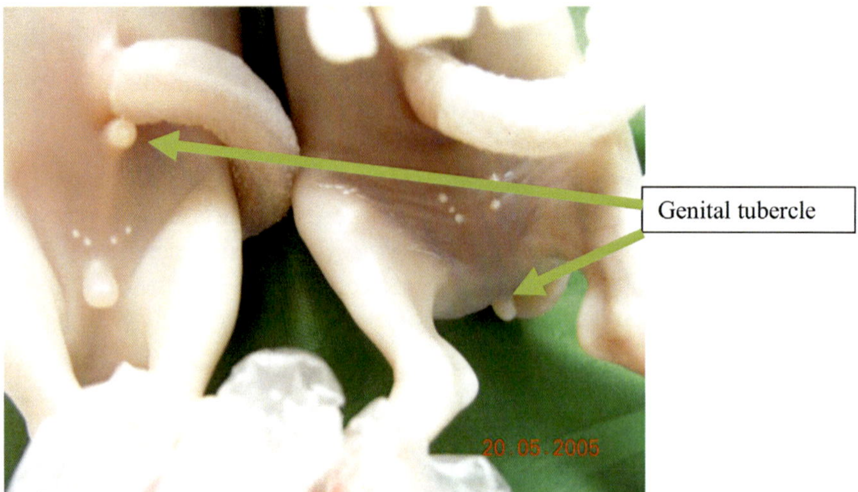

Photograph-7 is showing male fetus on left and female fetus on right side. See the position of genital tubercle close to umbilicus in male and close to tail in female. The male has irregular, while female has proper line up of mammary gland. In male scrotum can also be imaged in ultrasonography.

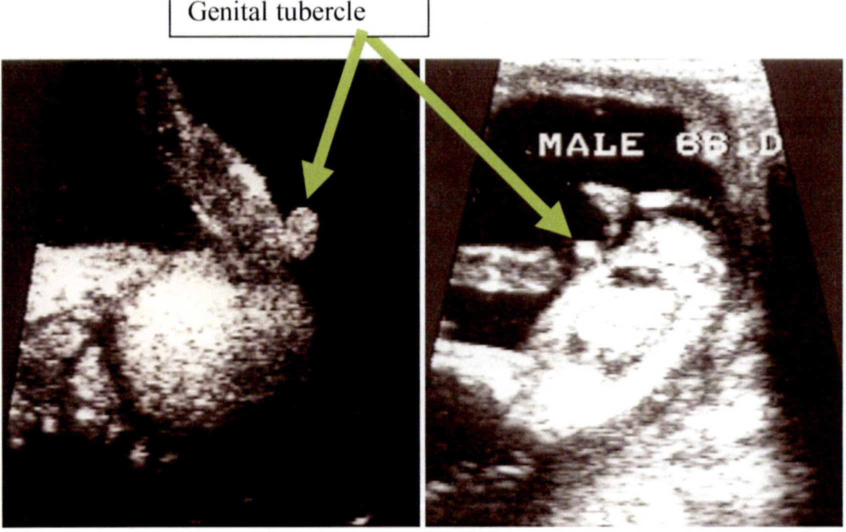

Fig. 18 & 19: are showing position of genital tubercle close to umbilicus (male fetus)

It is recognized as a prominent bilobular structure in the vicinity of the umbilical cord in males and the tail in females. Correct diagnosis can be made between day 55-100 post breeding in buffaloes.

Anoestrus and Silent Oestrus

The classification of animals that are non-cyclic (true anoestrus) or are cyclic but estrus is not detected (silent oestrus), is normally based on observing the behavioral signs of oestrus by the animal owner or rectal palpation of ovaries for the presence of CL by veterinarian. However, both methods do not provide reliable information as discussed earlier. Transrectal ultrasonic scanning can provide accurate information about the status of the CL and cyclicity of the animal. These observations can be helpful in predicting the day of oestrus and ovulation.

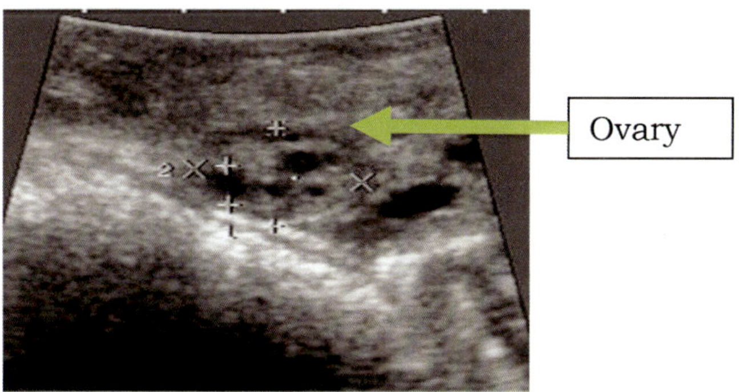

Fig. 20: Showing follicles in anoestrus buffalo. The size of ovary was 1.67 cm x 1.15 cm, while size of most of follicles was around 4.6 mm.

Oestrus Induction and Ovulation Synchronization

Ultrasonic scanning of ovaries provides useful information for rational treatment of silent estrus and anoestrus conditions in buffaloes. Buffaloes detected with a mature CL in the ovary, can be administered with single or double injections of PGF2α, 11 days apart. This is followed by insemination at observed estrus, usually occurring within 2-5 days post-injection. Acyclic buffaloes (no CL - True anoestrus: delayed puberty, post partum anoestrus, summer anoestrus), on the other hand, essentially require progesterone priming to induce fertile ovulatory oestrus. It has been revealed by ultrasonic examinations that wave like pattern of ovarian follicular growth continues even during anoestrus period and diameter of the largest dominant follicle (DF) may attain a size equivalent to that of the preovulatory follicle (Sharma *et al.*, 2004). This implies that anovulatory anoestrus condition is due to failure of the DF to ovulate rather than its absence. In anoestrus buffaloes, ultrasonography revealed that the DF undergoes atresia rather than ovulation, possibly due to failure of appropriate preovulatory LH surge. Exogenous GnRH caused ovulation / luteinization of large DFs, leading to subsequent formation of CL which can be readily identified with ultrasonography (Sharma *et al.*, 2004).

Ultrasonography in Embryo Transfer Program

In ET program ultrasound can be used to evaluate superovulatory response and devising suitable strategies which could enhance ovulation rate in donors. The response to a given quantity of gonadotrophin is extremely variable both between and within individuals. Ultrasonic examination of recipient buffalo for the presence of CL on the day of embryo transfer (Day 5/6) is very helpful for its correct diagnosis. The embryo must be transferred into the uterine horn, ipsilateral to the side ovary bearing corpus luteum. Accurate identification of early CL by rectal palpation is extremely difficult especially in buffaloes.

Ultrasound Guided Ovum Pick up and IVF Technique

Considerable progress has been made in harvesting oocytes from slaughter house ovaries and creating suitable media for in vitro maturation, fertilization and subsequent embryo culture. However, limitation in this technology has been the pedigree status of oocyte collected from abattoir. Curiously, almost 99% population of the follicles present in the ovary undergoes atresia and only less than 1% of it is used in the active reproductive phase of buffaloes. Therefore, use of transvaginal ultrasound guided oocyte aspiration technique in live animals is a better alternative to trap these unutilized oocytes and produce elite embryos of known genetic make up using IVM-IVF technology.

Apart from oocyte aspiration, ultrasound guided puncture technique can also be used for aspiration of fetal fluids for sex determination, biochemical analysis and hormonal estimation; sampling of uterine contents for diagnostic purposes and injection of substances into the ovaries, follicles, corpus luteum and uterus for research purposes.

8

Basic Ultrasonography for Field Veterinarians Assessing Bovine Reproductive System

The recent availability and use of ultrasonography in veterinary science has revolutionized the study of bovine reproduction. The information- gathering capabilities of this technology are much higher than those of conventional rectal palpation. Ultrasonic imaging has provided new insights into the complex biology of various reproductive processes such as folliculogenesis and early fetal development. With the pressure on agricultural land, due to ever increasing human population, the grazing grounds for animals are fast squeezing and as a result more and more animals are being stall-fed. This practice has adversely affected the availability of essential nutrients to the animals and their reproductive efficiency. The problem is graverin small states, where pressure on land is more to accommodate the expanding developmental and commercial activities of the national capital situated next door. The routine applications of ultrasonography include evaluation of animals for pregnancy, twinning, cyclic structures on ovaries, uterine and ovarian pathologies etc. The specialized but limited uses of this technology are transvaginal follicular aspiration for recovery of oocytes and as a complimentary technology to embryo transfer. To derive maximum benefits of this technology, it should be integrated with the conventional rectal palpation procedures for accurate and prompt diagnosis. As the animal owners are becoming more knowledgeable, demanding and concerned about their profits, there is a very low tolerance to poor services including delayed or incorrect diagnosis. Under these circumstances, the need to adopt ultrasonic imaging in bovine reproduction for providing better services is inevitable.

Principles and Basic Techniques

Ultrasonography is a non-invasive and non- disruptive tool that provides access to the internal reproductive organs. So far, skillful rectal palpation has been the only alternative for examination of the internal genitalia. This recent technology utilizes high frequency sound waves (above the audible limit of

20,000 cycles per second) to produce images of the tissues and internal organs. Ultrasonic imaging is one of the most revolutionary advances in animal reproduction having a variety of clinical and research applications. Unlike wireless waves, the ultrasound waves cannot travel across a vacuum. These are also difficult to propagate through air or from air into liquid/ solid or vice versa. The ultrasonic waves need some sort of transmission medium and are thus confined to liquids and solids. Most units used for ultrasonic imaging operate at a frequency of two to seven million cycles per second (mega hertz).

There are many ways to produce ultrasound, but for medical purpose, the use of crystals as ultrasonic generators is universal and dates back to the last century when the Piezo-electric effect was discovered. Piezo is derived from Greek word 'Piezein' meaning to press. Piezo crystal produces a potential difference across its opposite faces when under mechanical stress (pressure). For modern ultrasonography, the sound waves are produced by vibrations of specialized crystals (piezo electric crystals) housed in a transducer. Vibrations of crystal are produced in response to pulses of electric current. The crystal may be pulsed 1000 to 2000 times each second and the crystal will vibrate at a rate of 2 to 7 million times per second during individual pulse. The sound waves are directed towards the organ or tissue of interest by moving or changing the angle of the probe. The echogenecity of an organ or tissue depends on its capacity to reflect sound waves or acoustical impedance. The greater the density, the stronger the echo it will return. A proportion of ultrasound waves are received back from the tissue/ organ to the transducer, which could be used for interpretive evaluation. The transducer, therefore, acts as both the sender and the receiver of sound waves. One of the methods for evaluation is to make visual appraisal of echoed sound waves by using a cathode ray tube (Monitor) to trace the amplitude (A- mode) of the echoes produced as a function of time or distance. Moreover, this mode will measure echoes as waves and does not portray morphological information. A second method of evaluation converts the echoes to various intensities of brightness (B- mode) as a series of spots. When enough of these dots are present and there are adequate differences in brightness, an image can be formed on a display screen that allows a two-dimensional view of the tissue directly under the transducer. The brightness of the dots on monitor is projected in various shades of grey much like a black and white photography. If the B- mode display is moved over a recording device at a constant rate, it will record any motion of the tissue reflecting the echo, hence the term M-mode. The echoes appear as varying shades of grey (black to white) depending upon the echogenecity of the structures involved. For example, liquids are non-echogenic (do not reflect sound waves) and appear as black on screen whereas dense tissues such as bone are echogenic (reflect a greater proportion of sound waves) and look as light grey or white.

Scanners/ Transducers

Transducers of varying shapes and frequencies are available in the market. The frequency of a probe is determined by the number of times its crystal expands and contracts per second. In general, the higher the frequency of sound waves, more the image resolution, and the smaller the depth of area covered. A majority of the ultrasound scanners currently being used by animal reproduction biologists are B- mode (brightness modality) real time scanners, where as the brightness is proportional to the amplitude of the reflected echoes returning to the transducer. Real time indicates the ability to image moving objects e.g. foetal movement, heart beat etc. Ultrasound scanners that are available in the market vary greatly in their capabilities and additional provisions such as EGG, quantitative measurement, freeze frame memory, frame rate, focusing method, zoom magnification and photographic capabilities. The commonly used transducers in large animals have frequencies of 3.5, 5.0 and 7.5 MHz. The structures lying close to transducer (such as ovarian structures) are best examined with a probe of higher frequency i.e. 5.0 or 7.5 MHz. In contrast, the comparative large structures such as fetus and uterus that are located far away from the transducer are best suited for study with a 3.5 MHz transducer, since the depth of penetration and not the resolution of the image is important in such cases.

The Major Types of Transducers Available are Linear Array and Sector

Linear array: It makes use of a probe with an array of transducer elements. It forms a rectangular image on the monitor screen. The curved array is similar to the linear array technology, but the array of transducers is curved. This technology offers the possibility to focus at four different paths. Thus, the linear/curved array use advanced technology and makes them the probes of choice for many diagnostic applications.

Sector technology: The annular array (Sector) technology provides a very high image quality due to large size and round shape of crystals, which are divided into concentric rings, and forms a symmetrical beam resulting in an equally high lateral and transverse resolution. The large size of crystal makes it possible to obtain a very good signal to noise ratio resulting in a very high sensitivity. It is the most suitable technology for endo-cavity probes.

Ultrasound Equipment

The most equipments consist of a console (basic) unit possessing the electronics, controls and a screen to view images and a transducer, in addition to voltage stabilizer and accessories. The most useful transducer for routine reproductive examination in bovines is a linear-array, real-time, B-mode of 5.0 MHz. Small, developing follicles on ovaries are best appreciated with a 7.5 MHz transducer. However, a sector transducer is best suited for specialized applications such as transvaginal ovum pick-up. Most ultra-sound scanners are compatible with probes of different frequencies and shape.

Preparation of Animals

As the reproductive tract lies in close proximity to rectal wall, the transrectal ultrasonography is the most common approach to examination of genitalia in large animals. In general, the preparation and precautions are similar to those for rectal palpation. The animal should be properly restrained .The monitor screen should be at eye level with subdued external light. A coupling medium such as mineral oil or carboxymethyl cellulose (gel) serves both as lubricant and to provide contact between the rectal wall and the probe. All air bubbles need to be excluded as air is highly echogenic and reflects almost 100% of the ultrasound waves preventing screening of tissues underneath.

Imaging of Bovine Reproductive Organs

Ovaries

Ovarian follicles develop from the pool of nongrowing, primordial follicles present in the ovary since birth. They are fluid-filled structures surrounded by layers of granulose cells (inner) and thecal cells (outer). Such follicles appear as black circular structures surrounded by echogenic ovarian stroma.

The corpus luteum (CL) forms after ovulation and is a source of progesterone. CL appear as a distinctly echogenic area within the ovarian tissue. It may look like solid tissue mass and may also contain fluid-filled cavity. It needs to be differentiated from ovarian cysts. It may be difficult to distinguish between the growing and regressing corpora lutea.

Ovarian cysts are routinely diagnosed through rectal palpation as large fluid-filled structures. However, differentiation between different types of cysts (follicular / luteal) through rectal palpation is difficult even for a well-experienced veterinarian. Use of ultrasonography increases the accuracy of diagnosis. It may also assist to know if the cysts are single or multiples, their size and accompanying pathological changes in genital organs, if any. Ultrasound imaging may also be used to aspirate follicular cysts through transvaginal approach. It may also assist to know if the cysts are single or multiples, their size and accompanying pathological changes in genital organs, if any. Ultrasound imaging may also be used to aspirate follicular cysts through transvaginal approach.

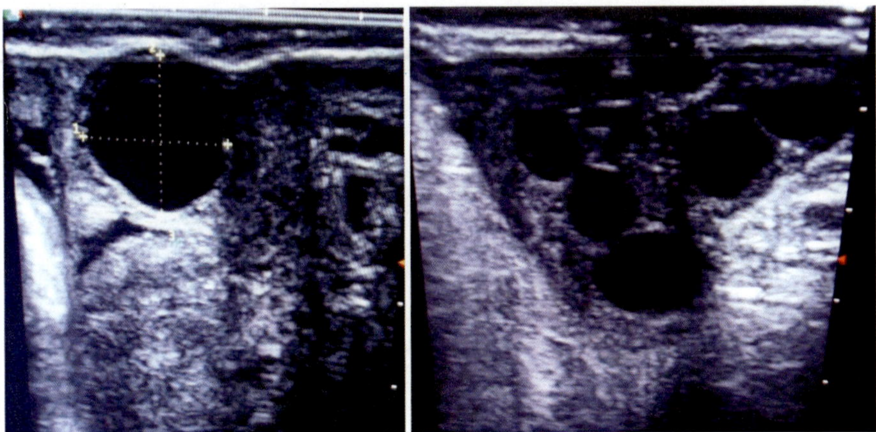

Fig. 21 & 22: is showing single (left) and multiple follicular cysts (right) in the buffaloes

Ovarian structures contain a lot of information, highly useful in diagnosing the reproductive status and in selecting the most appropriate intervention and therapy to increase the reproductive efficiency. As rectal palpation is poor in resolving different ovarian structures, using ultrasonic imaging for rapid and correct assessment remains need of the hour.

Uterus and Conceptus

Early pregnancy: The primacy objective for the use of ultrasound examination in animal reproduction is to assist in the early, accurate detection of pregnancy. The earliest stage of pregnancy at which a correct diagnosis could be made depends upon the quality of ultrasound unit, type of probe, species of animal and above all, the skill and expertise of the operator. The bovine embryo along with heartbeat can be visualized as early as 28 days post-service. It is more important and useful to diagnose non-pregnant rather than the pregnant animals at the earliest possible stage. Ultrasound is a much less invasive technique than the conventional rectal palpation. Ultrasonic imaging has also been used to identify early embryonic loss, twin pregnancy, developmental abnormalities and sex of fetus.

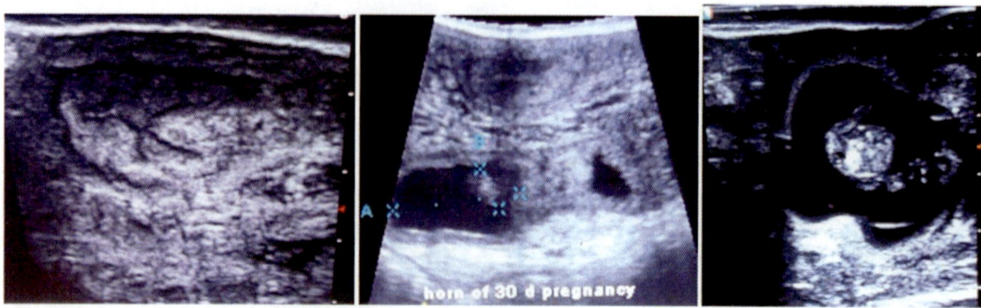

Fig. 23, 24 & 25 is showing section of non-pregnant horn **(left)**, 30 day pregnant **(middle)** and fetus encircled by amniotic sac **(right side)** in buffalo

Monitoring fetal gowth/well being: Ultrasonography is very useful in assessing proper growth of fetus. Fetal diamensions can be measured in length and width along with diameter of uterus. The length of uterus can not be measured but width of uterus can be measured upto certain stage of gestation.

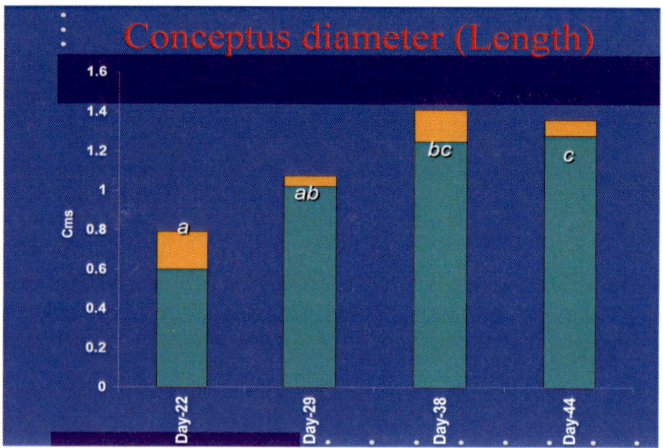

Graph 1: is showing gowth of fetus lengthwise fromday 22 to day 44 in Sahiwal cow from transrectal ultrasonography.

Anoestrus: After attaining puberty, a female animal should exhibit estrus every three weeks except during pregnancy or a short interval after calving. However, it does not happen in many cases. Today, anoestrus remains the principal deterrent in the process of optimization of reproductive potential of our indigenous cattle and buffaloes. The ultrasonic imaging of ovaries in anoestrus animals can be of great diagnostic value and help in taking proper remedial measures. Follicles as small as 2 to 3 mm in diameter can be easily identified using a good quality scanner and a high frequency probe. Small follicles are present on ovaries all the time. These follicles grow in waves and one of these becomes dominant and undergoes either ovulation or degeneration.

An animal, which is not cycling (true anoestrus) will not show any functional structure on any of the ovaries on imaging. Echotexture of genital organs changes according to hormonal status of animal. An experienced sonographer may easily differentiate the various stages of reproductive cycle from ultrasonic images of genitalia.

Apparent anoestrus: Some of the anoestrus animals have physiologically normal cycling pattern without overt heat signs. This may happen because of weak, silent suboestrus or missed estrus. It may also include animals that are either pregnant or had retained corpus luteum due to uterine pathology such as pyometra. Two gynaecological examinations at an interval of 10-15 days are required for differential diagnosis in such cases. This may be all the more difficult if the corpus luteum is embedded (not palpable, as is common in buffaloes). Ultrasonic imaging is of great help in such cases. A definite diagnosis is possible at the first ultrasonography itself. Pregnancy can be detected as early as 20 days. Thus, ultrasonography should save time as well as money.

Delayed ovulation: Delayed ovulation is the cause of repeat breeding in a good number of cases. A dynamic event like ovulation can be monitored by continuous ultrasonic evaluation of pre-ovulatory follicle. While the disappearance of a previously detected pre-ovulatory size follicle is the indication of recent ovulation, its continuous presence may indicate either delayed ovulation or follicular cyst. Similarly, ultrasonic imaging may help to identify repeat breeders because of anovulatory estrus or experiencing estrus in spite of being already pregnant.

Uterine pathology: Transrectal ultrasonic imaging has permitted assessment of changes in shape and size of the uterus during various reproductive states such as pregnancy, pyometra, hydrometra, fetal mummification, tumors, cysts etc. Echotexture helps in differential diagnosis and thereby the subsequent corrective measures. Further, this technique has also been used to non-invasively monitor uterine contractile activity, which is of prime importance for embryo mobility.

Ultrasonic imaging is also used to follow uterine involution following parturition. Uterine cysts appear as immobile, usually compartmentalized, non- echogenic structures with well-defined borders. Fetal remnants (bone) are ultrasonically detectable as bright echogenic structures within uterine lumen.

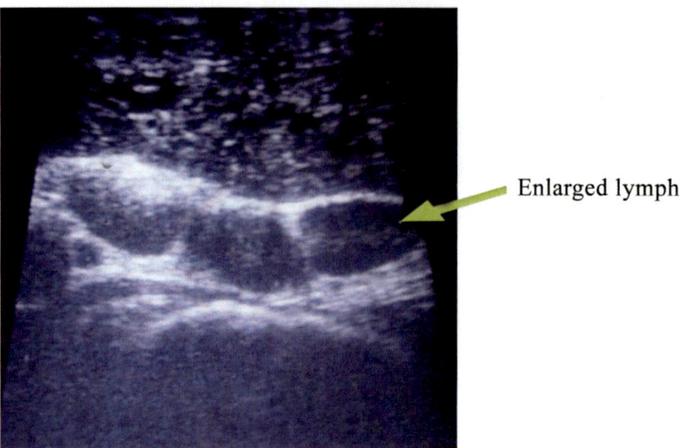

Fig. 26: is showing lymph adenopathy in the area adjuscent to uterus in a buffalo

Folliculogenesis/ Superovulation: Ultrasonography is useful in monitoring populations of follicles and their differentiation into various categories based on their antral diameter. Follicular population and their growth in wave pattern have been characterized and effects of various factors elucidated with the help of ultrasonic imaging. It has also made feasible to study intra-ovarian relationship among dominant and small (subordinate) follicles and corpus luteum.

Ultrasonography has also been used in bovines to evaluate follicular status before subjecting to superovulatory hormonal treatment and to predict and monitor the superovulatory response. Transvaginal ultrasonography has been successfully utilized for aspiration of follicular oocytes and in vitro genesis of embryos.

Miscellaneous uses: Ultrasonic imaging has been used for the characterization of twins, sex of the fetus and to know the precise timing of loss of fetus in infertile animals.

In conclusion, ultrasonography is a non-invasive, effective tool for directly evaluating reproductive organs, tissues and conceptus. However to derive maximum benefit, one requires good equipment and examination conditions in addition to keen observational skill. The major obstacles to success using this method in large animals include large abdominal size, the distance between the transducer and the target organ and the presence by gas- filled bowels between the sound source and the genital organs.

Tips for Ultrasound Examination

1. Ultrasound examination should be unhurried and methodical. It is quite possible to overlook features or misinterpret images if the operator is too hasty.
2. The examination area should be dark, as images are very difficult to see and follow if monitor is in excessive light.
3. The transducer's emitting surface is coated with a lubricant, so that no air pockets will occur between the transducer and the tissue being examined.
4. The ultra-sound unit should not be operated at a temperature exceeding 40 °C.
5. The probe should be carefully inspected after a fall or accidental damage. It may be dangerous to use a damaged transducer.
6. The probes may only be connected to or disconnected from the scanner when it is switched off. Ignoring this may cause severe damage to the scanner and / or probe.
7. The oil in the mechanical sector probes should never be replaced, only be filled up with special oil.
8. The probe should be cleaned after each examination as per supplier's guidelines.
9. Temperature of cleaning material should be in the range of 5-35 °C.

9

Estimation of Fetal Sex and Age by Ultrasonography

The ultrasonography is based on echo-principal i.e. reflection of sound waves. The ultrasound waves, commonly termed as ultrasound beam is generated from the pizzo-electric crystals, under electric stimulation. The sound beam travels and strike to the organs/tissues on its way and get either reflected or reabsorbed by these structures. The reflected beam is received back by the transducer, which are processed and displayed on the monitor of ultrasound scanner. The display of signal is in form of shades of grey scale i.e. from bright to black picture. If these sound beams strike to organ such as bone, then major part of the beam is reflected and it appears bright (white) on screen. If the sound waves strike to fluid, these get absorbed and are not reflected back, therefore, the image of fluid appears as black on the screen. Other organs appear in between black and bright white. In fact, the reflected sound beam is not proportionate to frequency and velocity of emitted ultrasound beam. The strength of ultrasound beam (grey scale) or brightness of the image depends upon the acoustic impedance of organ or tissue. In simple term acoustic impedance is the capacity of tissue or organ to reflect or absorb these sound beams that again depends upon density or composition of these structures. The reflection of sound beams also depends upon, the angle at which the ultrasound beam strikes the structure on its way. The distance of the organ from the transducer also affects image brightness.

Determination of Fetal Sex: The fetal sex is determined either on the basis of genital tubercle in bovine and equine or on the basis of scrotal sac in sheep and goat. In canine, imaging of intra-abdominal testis has been used, but with greater difficulties. The pattern of mammary glands in buffalo fetus has also been considered for sex determination. The bovine and equine, fetus sex can be determined around two months of gestational age.

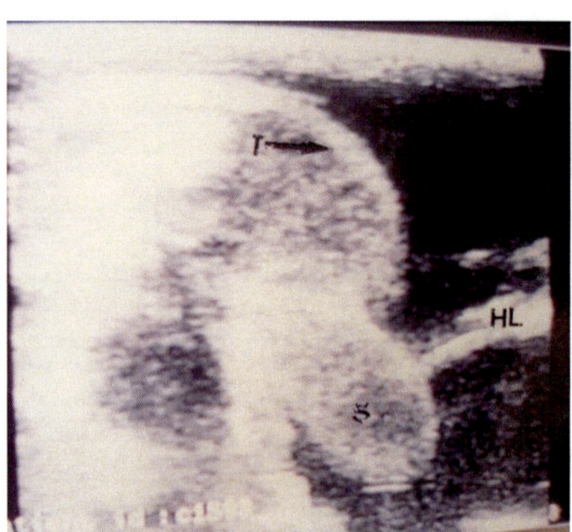

Fig. 27: is showing image of scrotum of ram lamb in womb on day 75

Importance of Estimation of Gestational Age of Fetus

The gestational age of fetus is commonly assessed in human obstetrical clinics. It is considered utmost important for determining expected date of delivery and parameters related to fetus. On the other hand, in veterinary practice estimation of gestational age (EGA) is not in common practice. Only few research studies have given various parameters for EGA. One reason might be non-availability of normal basal data to immediately know gestational age. In human, several formulae have been given on the basis of fetal dimensions. These dimensions are given for early, mid and late pregnancy. On that basis, one can measure parameter(s) and fit into formulae and get the estimation of age. These estimations are not 100% accurate, but in the range of 3 to 10 days variation, In veterinary practice EGA has been studied in equines. Few formulae have been suggested for EGA in cattle, sheep, goat and canine. These formulae are given along with delivery of ultrasound machine. One can take help of these formulae for EGA.

Fetal Parameters that can be used for EGA

Gestational Sac: Earliest fetal parameter for EGA is diameters of gestational sac (GSD) or circumference of gestational sac (GSC). In cows, buffaloes, sheep and goat, the gestational sac expands very early in longitudinal way, therefore, it is not possible to measure GC, but width of gestational sac (GS) is helpful. In mare, sow and bitch, the GS is round in early stage and its dimensions are useful for EGA.

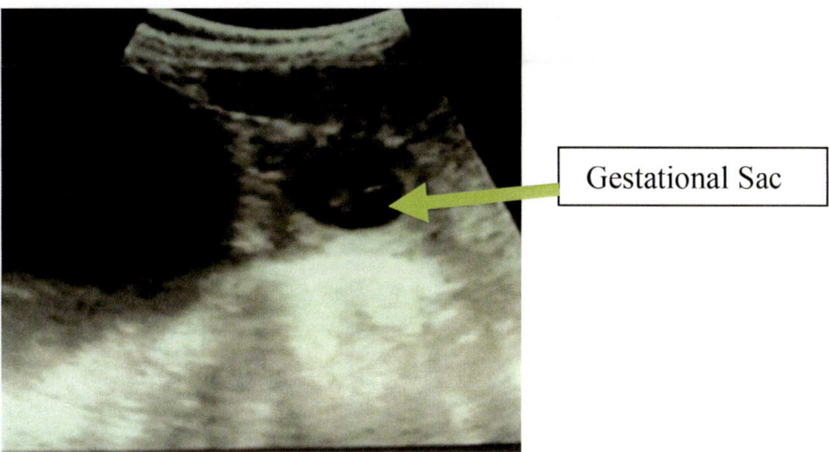

Fig. 25: is showing gestational sac with fetus in a bitch on day 20

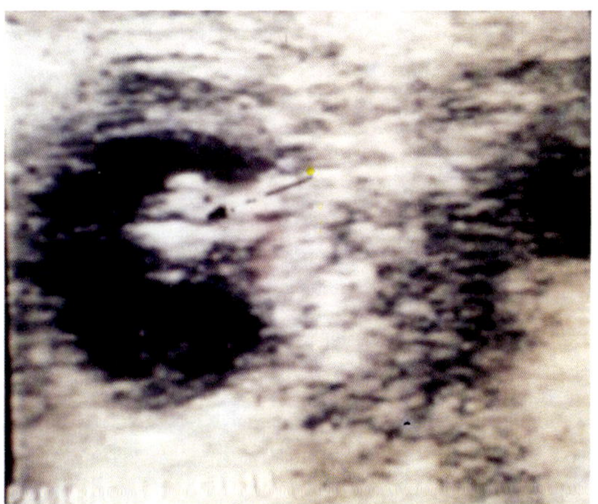

Fig. 28: is showing gestation sac (black pocket) with fetus (white) in a Sahiwal cow

In animals measurement of GS is upto a limited period. While doing trans-rectal ultrasonography, a linear transducer is used in large animals. This transducer has a limited length and width. It provides measurements of GS as per its own length and width. Therefore, this parameter has utility up to very early period in cows and buffaloes. This is very important parameter in mare, where the GS moves from one place to another. On this basis, it is possible to diagnose pregnancy and estimate gestational age.

Previously in cows and buffaloes, pregnancy was diagnosed on the basis of appearance of conceptus proper (day 28), but now on the basis of gestational sac, it is possible to diagnose pregnancy earlier than that (day 22-23).

Crown Rump Length (CRL)

Measurement of crown rump length is considered as important for EGA in early fetus. In the initial stages fetus in bovine and equine and other species undergoes several shapes. Therefore, this parameter should be measured as soon as fetal head is identified. In fact, the fetal body in cows and buffaloes shows tow clear division around 32 days. One round part has appearance like head i.e. it is round and smaller than others. The second part is body of fetus and is also referred as trunk in some texts. The full shape of fetus in cows and buffaloes is around day 45. This is the proper time of measuring CRL. The measurement is taken from pole to tale. This parameter is should be taken carefully, particularly when fetus is comma shape i.e. in bend form. The CRL looses its accuracy, once fetus grows beyond single-frame image. In bovine, it has been found alright till 55 days of gestation. Later on full fetus could not be taken in single frame. The CRL provides estimation of gestational age within error of 3 to 5 days.

Combination of mean diameter of gestational sac and crown-rump length are used to establish fetal age. Both parameters are useful because each measures a different aspect of the first-trimester pregnancy and may be used at different times during the first trimester.

Measurements of Head: The cranial cavity is clearly seen in ultrasonography. The fetal head bones encircling cranial cavity are also easily seen. Therefore, head parameters such as Bi-parietal diameter (BPD) and head circumference (HC) are measured routinely in human. In veterinary practice, this has been found one of best parameter for estimation of gestational age. In our observation, there was incremental increase in BPD between day 39 to day 90 in fetus of Sahiwal cow. The fetal head diameter (BPD) increased from approx. 1.0 cm at day 39 to 5.0 cm at 90 days of gestation.

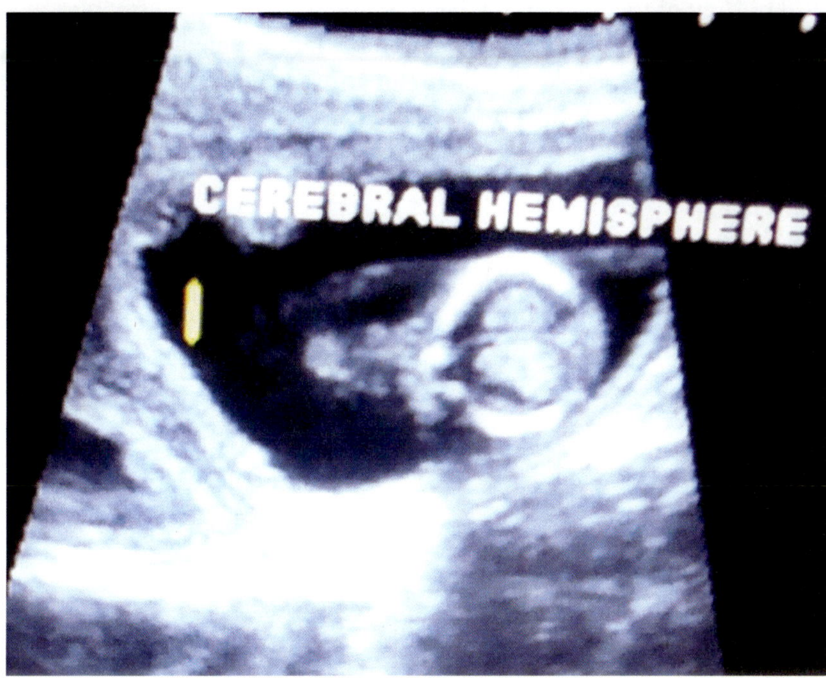

Fig. 29: is showing section of head, good for biparietal diameter to assess age of fetus in a buffalo.

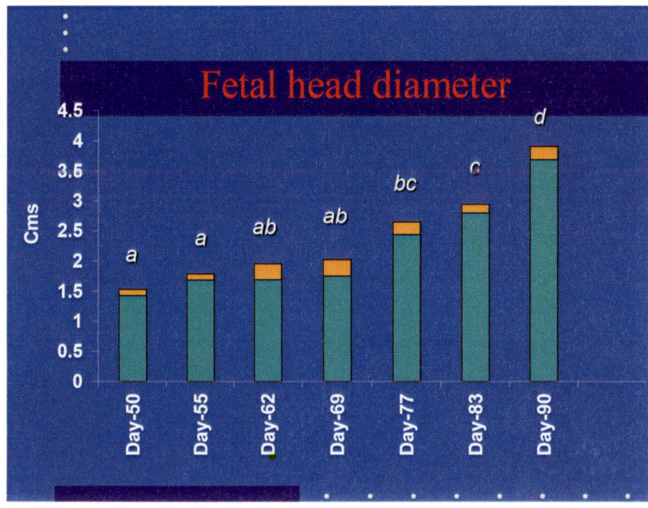

Graph- 2: is showing Fetal head diameter in womb of a Sahiwal cow

Abdominal Circumference (AC) or Trunk Dimensions: AC is obtained in the transaxial view of the fetal abdomen,at the level of the fetal liver, using the umbilical portion of the left portal vein as a landmark, The fetal stomach is at the same level, which is slightly caudad to the fetal heart and cephalad to the kidneys. During early fetal development, the trunk length and width can also be measured as most of machine used in the past in veterinary sciences were not providing good images for AC. The AC is less accurate than BPD.

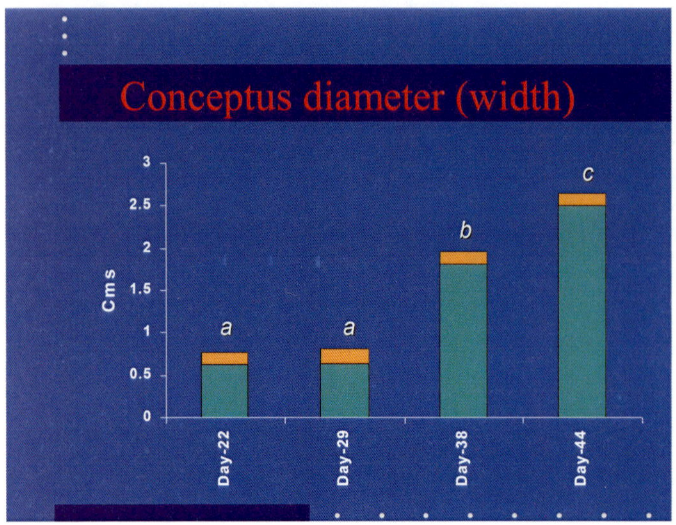

Graph- 3: is showing trunk dimension of fetus in Sahiwal cow from day 22 to 44

Ratio of HC/AC: The ratio of head circumference (HC) and abdominal circumference (AC) has been used to assess abnormality of fetal development. If this ratio is higher, then there are chances of mega head abnormalities. If ratio is less, then less development of fetal head (micro-cephaly is suspected.

Femur Length: The femur length is measured in advanced pregnancy. Data is available for human fetus. There are limited information on femur length measurement in farm animals. The fetus is mobile and femur length should be measured when fetus is at rest. For measurement of femur length, the ultrasound beam should be perpendicular to femur. The distal femoral epiphysis should not be included, when measuring femur length. There might be over- measurement during far-end imaging of femur. On the other hand, there might be under-estimate of femur length, when one fails to visualie entire length of femur.

FL/BPD Ratio: Theratiofemur length and BPD is also considered for EGA. Abnormal ratio is an indicator of pathologic entities, microcephaly (FL/ BPD abnormally high) and hydrocephalus or short-limb dysplasia (FL/BPD abnormally low). Abnormal ratio is an indicator of pathologic entities,microcephaly (FL/BPD abnormally high) or hydrocephalus or short-limb dysplasia (FL/BPD abnormally low).

Eye Ball Diameter: Eye ball diameter has been found positively correlated with gestational age. There are several observations of measuring eye ball diameter from trans abdominal scanning in pony mare. This is good parameter in advanced stage of gestation.

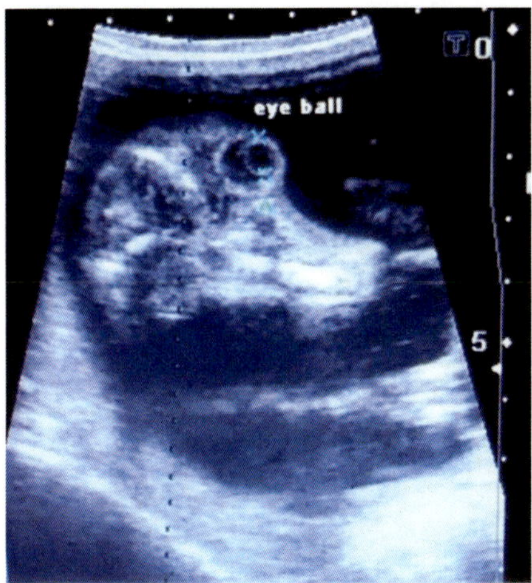

Fig. 30 is showing eye ball diameter in a buffalo calf.

Placentome diameter: In sheep, goat and cow, this parameter has been studied and found to be best correlated with gestational age.

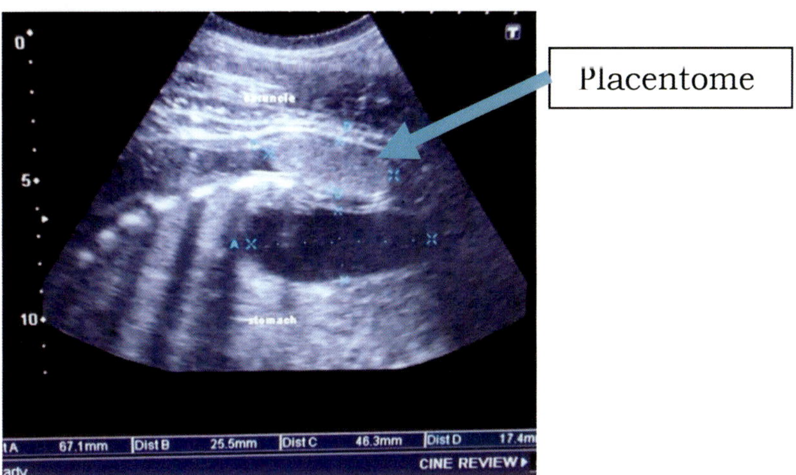

Fig. 31: is showing diameter of placentome in a buffalo.

Umbilical Diameter: The image of umbilicus can be easily obtained in cows and buffaloes. The diameter also changes with gestation stage. Some ultrasonographer advised to use diameter and circumference of umbilicus or its vessels for estimation of gestational age, however it is not very popular.

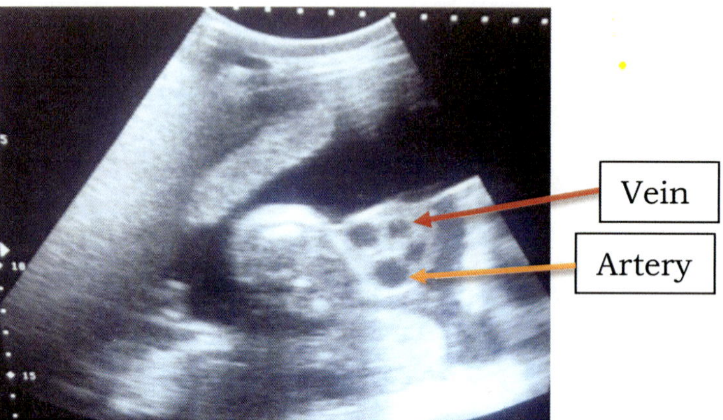

Fig. 32: is showing cross section of umbilicus in a buffalo during trans- abdominal ultrasonography. The veins have smaller diameters than arteries.

Gasa: Growth adjusted sonographic age (GASA). This is done taking several measurement of a parameter and growth rate is adjusted.

Combined Parameters: combined parameters or ratios are better indicator of gestational age. If one parameter estimates extremely high or low, then fetus should be thoroughly investigated.

10

Use of Transrectal Ultrasonography in Caprine Reproduction

Ultrasonography in goat is becoming popular in the field of reproduction and replacing the other old-age techniques, viz. abdominal ballotment, recto- abdominal probes, X-rays and radioimmuno-assays. The reasons are numerous such as unsatisfactory results, hazardous effects to animal, operator or environment, high cost, long delay in diagnostic results and requirement of sophisticated labs. The ultrasound system has low operating cost and does not need special labs as it can be done at farm itself. Ultrasonography is considered safe to the animal and the operator and provides quick and reliable results. The popular opinion about ultrasonography is that it is noninvasive imaging system and is rapid, safe and practical means of diagnosis of pregnancy and reproductive disorders in sheep and goat. The B-mode real time ultrasonographic system particularly portable system is popular among researcher and practitioners. In goat, B-mode real time ultrasonography was used in 1971, using 5 MHz frequency transducer. A transducer of switchable frequency from 7.5 to 5.0 is more popular in small ruminant research and practice.

Transrectal Procedure

Following are some of the steps to obtain good quality image

1. ***Evacuation of the rectum:*** The rectum should be evacuated by removing fecal material using two fingers i.e. fore-finger and middle finger.

2. ***Filling of rectum with gel:*** The ultrasonic gel should be infused in the rectum for better image.

3. ***Selection of transducer:*** A linear array transducer of switch-able frequencies is advised. The commonly used frequency in sheep is 5.0 MHz, however, 7.5 MHz is useful to detect small follicles and detail of other structures.

4. ***Preparation of transducer:*** The transducer should be prepared for rectal manipulation. Since, hand can not be inserted into the rectum, therefore,

the transducer should be attached to either a glass rod or to a PVC pipe with the help of a good quality adhesive tape. This assembly will be helpful in the manipulation of the transducer from outside the rectum.

5. ***Setting of machine:*** The machine should be set up at eye level to view the image properly. Machine should be connected to the power outlet. Make proper connections to the printer and VCR. In the beginning, the machine should be set for proper contrast, brightness, near-gain, far-gain and overall-gain. In most of the machines, there is a step-bar either on sides or top or bottom of the machine. This bar has different contrast steps. The machine should be set in such a manner that all the steps on the bar should be clearly visible.

6. ***Restraining of the animal:*** The animals should be put in the animal-chute otherwise lightly restrained by an attendant in standing position in an animal- pan.

7. ***Insertion of transducer into rectum:*** The transducer assembly is generally of larger diameter and animal feels pain at insertion of transducer. Therefore, lubrication should be liberal both into rectum and on the transducer. The transducer should be inserted full length (about 6 inches) and gently moved along the rectal floor and is rotated on about 50 - 60° on either side of midline.

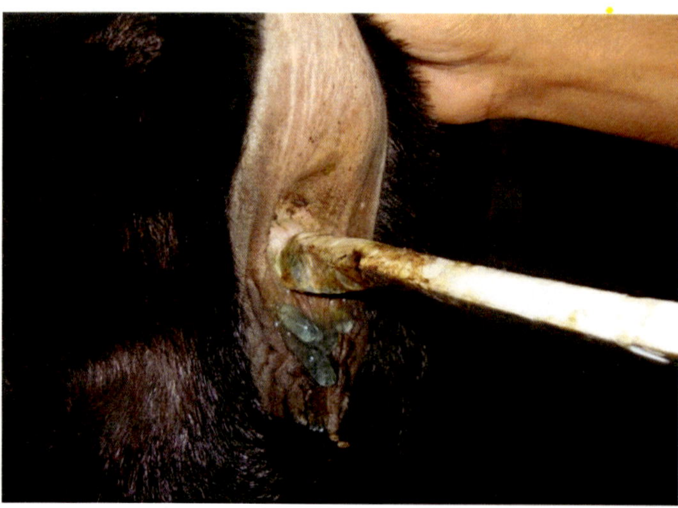

Photograph-8: is showing insertion of assembled probe into rectum of a goat

8. ***Landmark and identification of pelvic organs:*** The urinary bladder is taken as landmark, which is easily identified as anechoic (black). The tubular track is located on dorsal surface or either side of urinary bladder.

Identification of reproductive tract: The images of reproductive tract can be identified based on the experience. For the beginners, it is advised to practice on slaughterhouse material in a water-tub. The images of tubular reproductive tract should be taken in longitudinal and cross section. The images of ovaries are better identified during follicular phase. On the whole animal, the linear transducer should be pressed against the rectal mucosa and be manipulated from outside. The attached PVC pipe should be marked with a marker pen. The mark on pipe should be in the line with active phase of transducer so that it can be known from outside, where the active phase of transducer is. First, the images should be taken in longitudinal direction and then in cross section. In the beginning for location and scanning of the ovaries, animals should be given small dose of FSH or scan during follicular phase, as it is easy to identify the ovaries during this phase. The urinary bladder appears homogenously non-echoic i.e. completer black. The non-pregnant uterus should be imaged around bladder, generally lateral to bladder and ventral to rectum. The transducer should be rotated in angular fashion to obtain images of sections of uterine horns. The sections might be cross sections or oblique sections. For scanning of uterus, it takes about 5 minutes to complete entire procedure, provided one easily identifies the images. One should move transducer slowly to get clear images. Fast-movements may lead to missing of important structures.

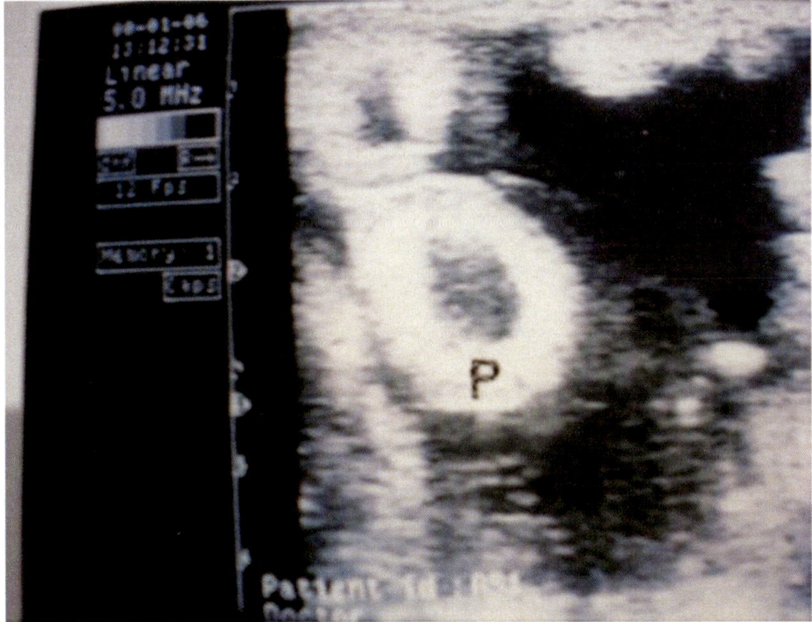

Fig. 33: is showing placentome **(P)** in a pregnant Black Bengal goat (Kumar *et al.*, 2005)

Trans-Abdominal Procedure

Trans-abdominal procedure is not very popular as it is difficult to obtain images due to obstruction by digestive organs, rumen on left and intestines on right side. The distant location of early pregnant uterus is another hurdle to obtain good images of fetus. This is good in advanced pregnancy particularly after three months of age. The goats to be scanned should be taken out of flock and kept near the ultrasound system. For transabdominal ultrasonography from flank region, shaving of flank, particularly posterior and lower flank is required. It is advised to shave the animal on both sides, because in many instances, the scanning may be required on both sides. It is better to do ultrasonography in the morning, before taking animals for grazing. The animal should be kept in lateral recumbency for ultrasonography. In author's experience, the proceduce in standing position is more difficult to take image of fetus than in lateral recumbeny.

The transducer is placed over upper last portion of the left flank region. The transducer should be moved left- right or up-down to find an area from where ultrasonic beam can pass through. It is good to observe fetal activities and antenatal fetal organs. A transducer of 5.0 MHz is found good for this procedure.

3.5 MHz Frequency can also be used in Late Stages of Gestation

Positioning of transducer on lower abdomen may get images of advanced pregnant uterus; however the entry of ultrasonic beam is difficult from this point. In the recent past, there are some transabdominal ultrasonographic data available. It has been suggested to place the transducer between thigh and udder.

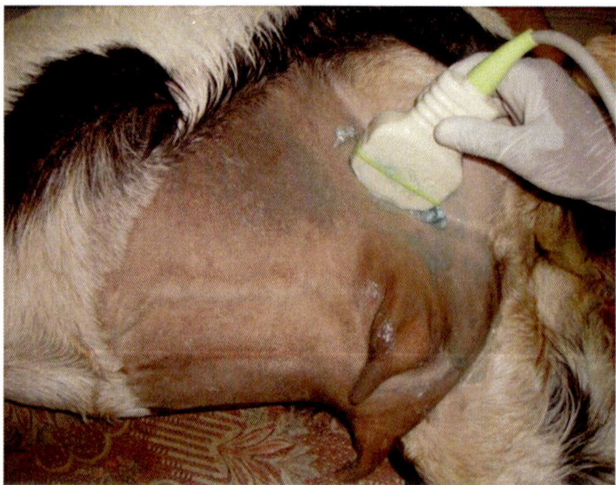

Photograph-9: is showing position of probe for transabdominal ultrasonography of a goat

Initial images from transabdominal ultrasonography can be obtained from day 35 onwards. Early observation before day 35 of pregnancy depends upon experience of operator, frequency of transducer and resolution of the machine.

Pregnancy Confirmation: Pregnancy diagnosis in goat is possible around end of third week, but confirmation is possible on middle of four week (from day 23 to 26). During this time, heartbeat of embryo can be easily identified. At this time, frequency of either 5.0 MHz or 7.5 MHz should be used.

Sex Determination: The diagnosis of sex is possible at the end of second month, but images of scrotum are obtained around day 70 onwards. It is easy to get images of scrotum in between the thighs.

Fetal Activities: The difference between goat and sheep regarding fetus development is in respect to position of fetus. Through per rectal scanning the head of fetus in goat is mostly downwards, while the sheep fetus keep head and mouth in easily identifiable position. To get images of head and mouth, one need to push abdomen upward or in later stages, the trans-abdominal images should be obtained. The fetal movements in goat are faster that in sheep fetus where movements are slow. In both species, the fetus, *in utero* shows sleeping mode, rumination like movement of mouth and engulfing of amnion. The fetus does all the activities in uterus that are observed after birth. Therefore, one can say that all the activities of life are learnt by a baby in the womb of mother.

Cotyledones: The cotyledons are seen very easily. From different angles they are seen either as full moon shape with a hypo-echoic center or half moon shape, concave in appearance.

Antenatal Fetal Organs: The stomach, liver, kidney and urinary bladder of the developing fetus are easily identifiable.

Variation in Heartbeat of Fetus: The heartbeat of the developing fetus is very fast in initial stage of development and then slows down as pregnancy advances.

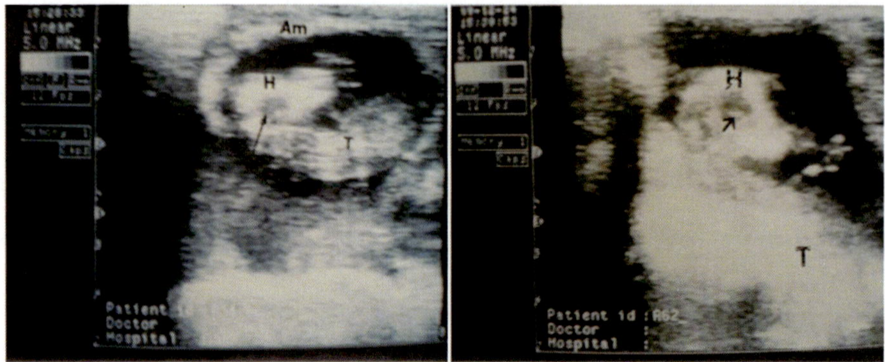

Fig. 34 & 35: are showing full foetus in a pregnant goat (Kumar *et al.* 2005)

Fetal Mortality: The fetal mortality has been diagnosed in one of our study, where the fetal fluid became cloudy and fetus becomes motionless without heartbeat. In one of the case in goat, early pregnancy was diagnosed and all pregnancy dimensions increased in next scanning. In third scanning, there was decrease in diameter of conceptus and heartbeat slowed down. In fourth scanning, there was no sign of fetus only small amount of fluid was observed that was also reabsorbed in due course of time.

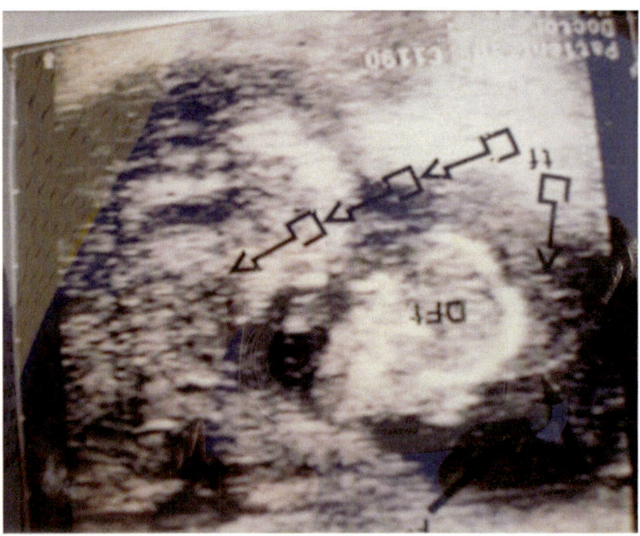

Fig. 36: is showing a dead fetus in utero in a goat. Fluid accumulation among muscles and joints.

Impending Fetal Death: This could be judged from slowing heart-beat and change in pixel-values of amniotic fluid.

From our experience in goat, it could be concluded that the in-utero fetal development including fetal organs can be studied using per rectal ultrasonography. The sub-cutaneous (abdominal) ultrasonography can be done in goat during advanced stage of gestation.

11

Ultrasonography of Reproductive Tract in Sheep

Ultrasonography in sheep is becoming popular in the field of reproduction and replacing the other old-age techniques, viz. abdominal ballotment, recto- abdominal probes, X-rays and radioimmuno-assays. The reasons are numerous such as unsatisfactory results, hazardous effects to animal, operator or environment, high cost, long delay in diagnostic results and requirement of sophisticated labs. The ultrasound system has low operating cost and does not need special labs as it can be done at farm itself. Ultrasonography is considered safe to sheep and the operator and provides quick and reliable results. The popular opinion about ultrasonography is that it is noninvasive imaging system and is rapid, safe and practical means of diagnosis of pregnancy in sheep. The B-mode real time ultrasonographic system particularly portable system is popular among sheep breeders. In sheep, B-mode real time ultrasonography was used in 1971, using 5 MHz frequency transducer. A transducer of switchable frequency between 7.5 to 5.0 is more popular in sheep practice.

Transrectal Procedure in Sheep

The procedure is similar to goat, but sheep remain quite during transrectal ultrasonography than goat. Following are some of the steps to obtain good quality image:

Evacuation of the rectum: The rectum should be evacuated by removing fecal material using two fingers i.e. fore-finger and middle finger.

Filling of rectum with gel: The ultrasonic gel should be infused in the rectum for better image.

Selection of transducer: A linear array transducer of switch-able frequencies is advised. The commonly used frequency in sheep is 5.0 MHz, however, 7.5 MHz is useful to detect small follicles and detail of other structures.

Preparation of transducer: The transducer should be prepared for rectal manipulation. Since, hand can not be inserted into the rectum, therefore, the transducer should be attached to either a glass rod or to a PVC pipe with the help of a good quality adhesive tape. This assembly will be helpful in the manipulation of the transducer from outside the rectum.

Setting of machine: The machine should be set up at eye level to view the image properly. Machine should be connected to the power outlet. Make proper connections to the printer and VCR. In the beginning, the machine should be set for proper contrast, brightness, near-gain, far-gain and overall-gain. In most of the machines, there is a step-bar either on sides or top or bottom of the machine. This bar has different contrast steps. The machine should be set in such a manner that all the steps on the bar should be clearly visible.

Restraining of the animal: The animals should be put in the animal-chute otherwise lightly restrained by an attendant in standing position is advised in an animal-pan.

Insertion of transducer into rectum: The transducer assembly is generally of larger diameter and animal feels pain at insertion of transducer. Therefore, lubrication should be liberal both into rectum and on the transducer. The transducer should be inserted full length (about 6 inches) and gently moved along the rectal floor and is rotated on about 50 - 60° on either side of midline. Landmark and identification of pelvic organs: The urinary bladder is taken as landmark, which is easily identified as anechoic (black). The tubular track is located on dorsal surface or either side of urinary bladder.

Identification of reproductive tract: The images of reproductive tract can be identified based on the experience. For the beginners, it is advised to practice on slaughterhouse material in a water-tub. The images of tubular reproductive tract should be taken in longitudinal and cross section. The images of ovaries are better identified during follicular phase. On the whole animal, the linear transducer should be pressed against the rectal mucosa and be manipulated from outside. The attached PVC pipe should be marked with a marker pen. The mark on pipe should be in the line with active phase of transducer so that it can be known from outside, where the active phase of transducer is. First, the images should be taken in longitudinal direction and then in cross section. In the beginning for location and scanning of the ovaries, animals should be given small dose of FSH or scan during follicular phase, as it is easy to identify the ovaries during this phase.

Pregnancy Diagnosis: Pregnancy diagnosis in sheep is possible around end of third week, but confirmation is possible on middle of four week (from day 23 to 26). During this time, heartbeat of embryo can be easily identified. At this time, frequency of either 5.0 MHz or 7.5 MHz should be used.

Diagnosis of sex: The diagnosis of sex is possible at the end of second month, but images of testes are obtained around day 75 onwards.

Transabdominal Ultrasonography in Sheep

For transabdominal ultrasonography, the sheep is kept in lateral recommency. The probe is placed between hindlimb and udder, close to limb. Any deviation towards udder may not generate image. For pregnancy diagnosis this approach is useful after 30 days of pregnancy. As pregnancy advances, the ultrasonography is easier, but there is interference of abdominal wool. Therefore, for scanning of sheep in advance pregnancy, proper shaving is required. One should be liberal in using gel, while doing transabdominal ultrasonography. Ovaries can not be approached from this approach. For imaging ovaries, transrectal procedure is better.

12

Ultrasonography for Reproductive Disorders in Bitch

Introduction

Ultrasonography is well-established technique in human sciences, particularly for diagnosis of pre-natal defects of fetus. In veterinary sciences, this technique is emerging an important tool. This tool is being employed in the small animal practices, particularly bitches for diagnosis of various reproductive disorders and estimation of gestational status. In the past, veterinarians had limited assistance for assessing reproductive status of this animal. Pregnancy diagnosis in a bitch was possible either with the use of X-rays or transabdominal palpation, but that too at an advanced stage of gestation. These methods could provide information either about presence of pups in vitro, or their number, but assessment of fetal viability was not accurately predicted with these traditional methods. With the use of X-rays, there was always risk of harmful effects on the health of mother and fetus. As far as reproductive disorders are concerned, most of them remained either undiagnosed due to the lack of proper access to the internal organs or were diagnosed with a guess. Diagnostic ultrasound is a valuable alternative imaging system that can provide more accurate information about the canine pregnancy and reproductive disorders in comparison to traditional methods. Ultrasonography is of many kinds such as A-mode, B-mode, M-mode and Doppler ultrasonography. For diagnostic purposes, B-mode ultrasonography is commonly employed. In this mode, various frequencies are available. In bitches, frequencies between 4.0 to 7.5 MHz is used for scanning of reproductive tract. Preparation of the animal: The ultrasonographic scanning in bitches is done from lower abdomen/pelvic area. In this area, there are hairs that interfere in obtaining good quality images; therefore shaving of this area with a sharp razor is advised. Most of the bitches follow instructions from the pet owner, but for precaution (to avoid biting), the mouth of the animal should be tied with a cotton rope or a bandage. Light sedation is not usually required, but may be used if animal is not co-operative or pet owner is not accompanying the pet.

Type of Transducer: For scanning of reproductive organs of a bitch, a sector scanner should be used. Either a linear sector scanner or a convex sector scanner can be selected as per the availability. If a scanner of switch-able frequency is available, use that, it will be helpful in getting desired magnification and depth area of the tissue/organ. A transducer having frequency of 5.0 and 7.5 MHz can serve the purpose.

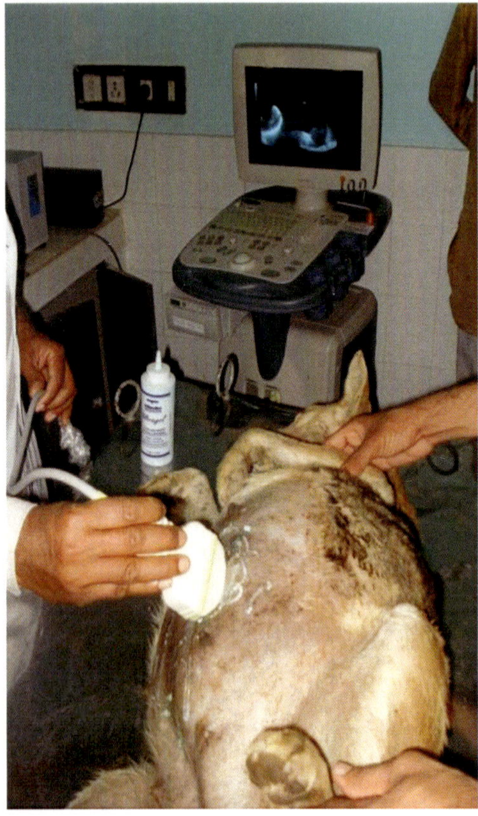

Photograph- 10: is showing control and scanning in dorsal position of a bitch

Ultrascan Gel: These water-based gels are available at a price ranging from Rs. 20 to 40. The quality of gel should be checked before making large purchases. Some gels are very thin and do not stick to the animal body; therefore are not useful. The gel should be of good quality that can serve the purpose of excluding air from the space between the transducer and animal skin.

Scan Area: The ultrasonography in the bitch is performed trans- abdominal/ trans-cutaneous. The scan area ranged from pubis to cranial of the umbilicus or beyond umbilicus in pregnant animal.

Procedure of Scanning

Position of Dog: Keep the animal in dorsal or lateral recumbency. One person should take care of forelimbs and neck, while other should hold hind limbs. Docile dog follows command of owner and may lie down accordingly, while other dogs need to control from hind and forelimbs.

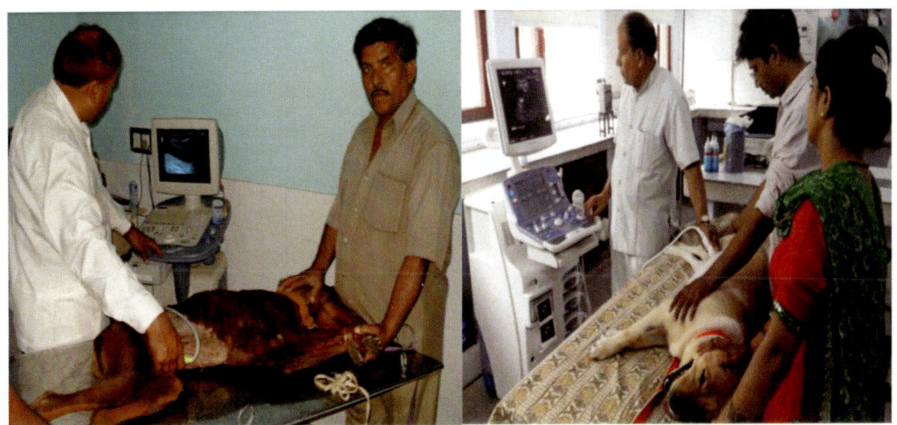

Photographs-11 & 12: are showing position and control of dogs for ultrasonography in lateral recumbency

The Monitor of Ultrasound Machine should be at Eye-level

Apply gel on transducer and animal body. The transducer is moved from caudal to cranial direction. The area of movement of transducer should be between and around two rows of mammary glands. The urinary bladder should be taken as landmark to locate reproductive organs.

The image of non-pregnant uterus can be obtained in the area between two rows of mammary glands, while during early pregnancy; images are obtained by moving transducer lateral to mammary glands.

In large bitches and pregnant animal, scanning can be performed in standing position, but in our experience, respiration and movement of animal interfere with scanning. There is also a problem of holding transducer in upside direction and one may loose contact easily, while moving transducer from one to another place.

Evaluation of Pregnancy Status

Observable signs of pregnancy in the bitch include weight gain, increased appetite, mammary enlargement and lactation. These signs occurs due to changes in endocrinological environment of hypothalamo-hypophysialgonadal axis that influences reproductive tract, but similar changes are also possible in pseudo-pregnant status. Real-time B-mode ultrasonography is an easy way for the differential diagnosis of normal pregnancy and pseudo pregnancy or other conditions in a bitch. Many investigators are of the opinion that pregnancy can be detected from 17-21 days after the surge of plasma luteinising hormone. The real-time ultrasonography is a reliable method for pregnancy diagnosis in bitches from 21-22 days of pregnancy and for litter-size estimation the most appropriate time is between day 25 and 30 of pregnancy. False positive diagnosis of pregnancies may occur, but rarely.

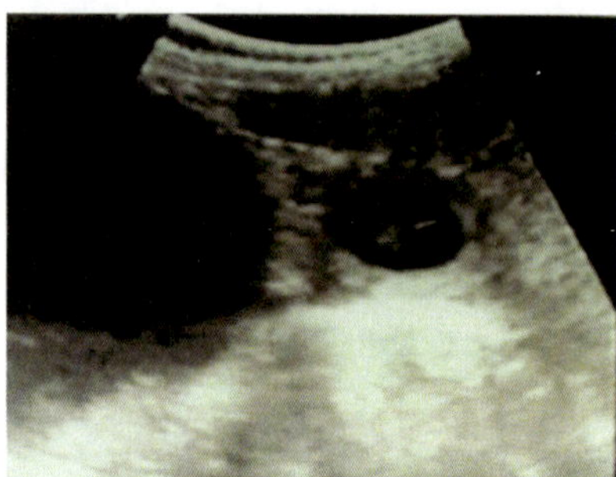

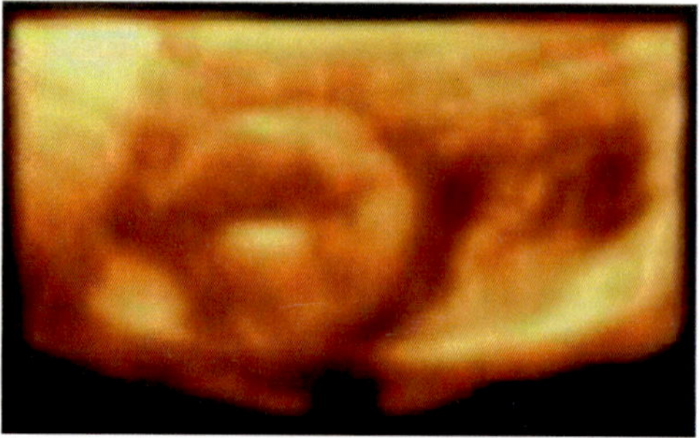

Fig. 37 & 38: are showing fetus on day 22 **(left)** and 3D image of fetus in sitting position on day 38 in bithches.

In addition to diagnosis of pregnancy, the ultrasound examination is also useful for monitoring embryonic and fetal well being on the basis of heartbeat and fetal daily/weekly growth. With advancing knowledge in this field, now it is possible to diagnose early pregnancy from day 18 onwards in Indian conditions, but one requires a good quality ultrasound machine. In our situations, where LH estimation is not always possible, there is a difficulty in ascertaining the exact date of pregnancy. In such cases, an estimation of stage of gestation is assessed on the basis of fetal development. Otherwise, the last date of mating is also taken as approximation for judging fetal age. Some parameters, regarding fetal development, are also published in literature for several breeds that can serve as database for the veterinarians. One should keep it as a reference to estimate fetal age in clinical practice.

Scanning on day 18 of gestation, the embryonic vesicle appears as spherical anechoic structures of around 2-mm diameter. It is actually the yolk sac contents that fill the chorionic cavity, which appears as the anechoic structure and is the first ultrasound confirmatory image of early pregnancy. The conceptus is initially somewhat spherical in nature, rapidly increases in size and may becoming oblate in appearance. On day 22 we can first recognize the fetal heart beat as rapid flickering in the center of the embryonic mass. The heartbeat ranges between 192 to 216. Accuracy of counting depends upon high long one can keep in focus the heart. During the same period, a 2nd fluid filled region adjacent to the embryo is also observed. The echoic images of developing allantois and the amnion may appear on 25th day. Amnion membrane is easily diagnosed in the fluid, while allantois need some experience.

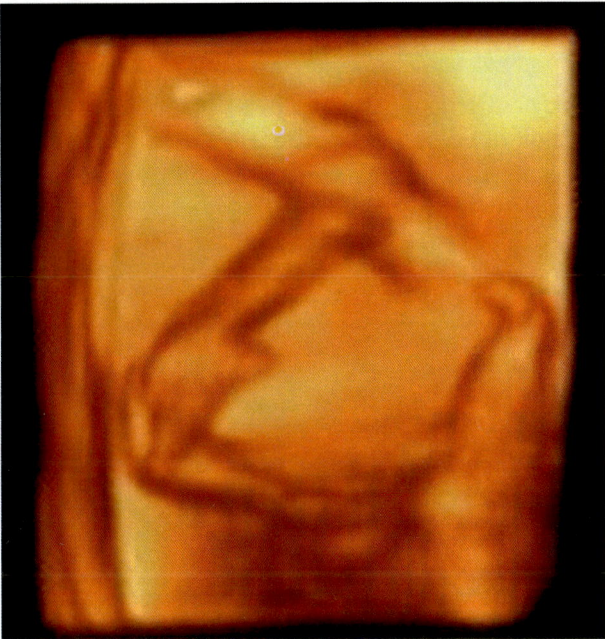

Fig. 39: is showing 3D view of Zonary placenta in a bitch, attached to fetus **(down right)** and uterus **(upper middle)**.

Fetal growth is more rapid in between 32nd and 55th day of gestation. Before this time, only the heart and focal anechoic region within the cranial pole can be appreciated. Marked differentiation of cylindrical fetus into its head, trunk, tail and umbilical cord is visible on 30th day. Hyper echoic fetal bones and heart valves can be appreciated on 40th day. Greater blood vessels are also seen during scanning at this time. In general from day 40th onwards only trunk diameter exceeds the head diameter. The hyper echoic regions of lungs and liver could be clearly distinguished approximately 20 days before parturition and fetal stomach can be imaged as an anechoic structure caudal to the liver around the same period itself. On careful slow scanning, fetal urinary bladder is also observed around this time. Hyper echogenicity of the fetal skeleton get enhanced during late pregnancy. With advancing gestation, the fetal kidneys can be imaged and fetal vasculature become more pronounced. In late pregnancy, intestinal loops may be detected. Around day 50, the images of bladder are easily obtained as a 2nd anechoic structure in caudal abdomen. One more, anechoic structure can be located within the liver, which is the gall bladder during the same stage.

Abnormalities of Pregnancy

Spontaneous Embryonic Resorption

If the embryonic death occurs within 25 days of gestation, then it gets resorbed in the uterus of the bitch. Cases of fetal death and later resorption have been recorded in this species. If death of one or two pups occurs, then continuation of pregnancy, even after resorption of such pups, occurs. In cases of embryonic resorption, the embryonic fluid volume reduces and changes in normal echogenic pattern are evident. There will be marked loss of the embryonic mass and absence of fetal heart beat, thickening and inward buldging of uterine wall with collapsed and reduced conceptus when compared to adjacent conceptus.

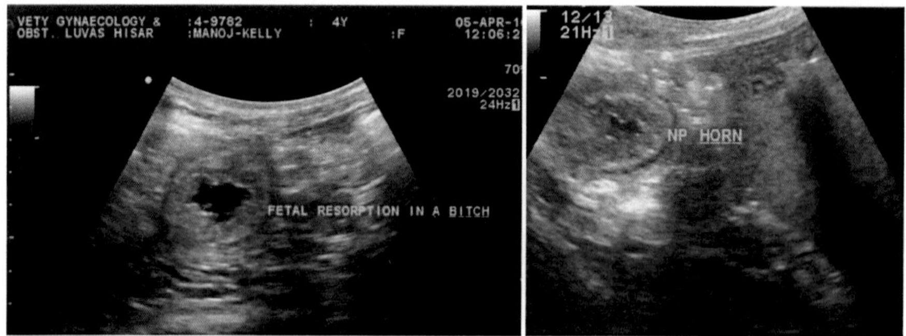

Fig. 40 & 41: are showing images from differnt parts of uterine horn in a bitch having fetal resorption. Fluid with some fetal remnants in left image and tiny amount of fluid in image on right sidein a bitch.

In cases of complete resorption of fetuses, uterine ultrasonographic picture will be similar to that of post partum uterus; it will appear as homogenous and moderately hypo echoic areas. The beginner ultrasonographer can predict the ensuing mortality if there is sudden slowing down of heartbeat or growth of the pup retarded in comparison to others or the fetal fluid turns turbid. The internal uterine diameter can also be used to predict normal and abnormal development of conceptus.

Fetal Abortion

During fetal death, the echogenicity of fetal fluid will be increased along with thickening of the uterine wall. If fetus is still inside, one should identify fetus proper and presence or absence of heartbeat. In later stages of fetal death, fetal fluid volume gets reduced and fetus appears as flecked because of fetal tissue breakdown. The investigator also reported that in such cases, the conceptual sac collapses along with thickening and inward buldging of uterine wall. If fetus is already expelled out or resorbed then, the uterine images will be just like that of non-pregnant status. There may be some amount of fluid in the uterus in late abortion cases.

Fetal Abnormalities

Common imaging regimes have a high success rate for detecting normal and abnormal fetuses. Enlargement of gestational sac and herniation of small intestine has been reported in pups and a case of hydrops fetalis has been diagnosed in which fetal fluid was identified within fetal thorax, pericardial sac and subcutaneously.

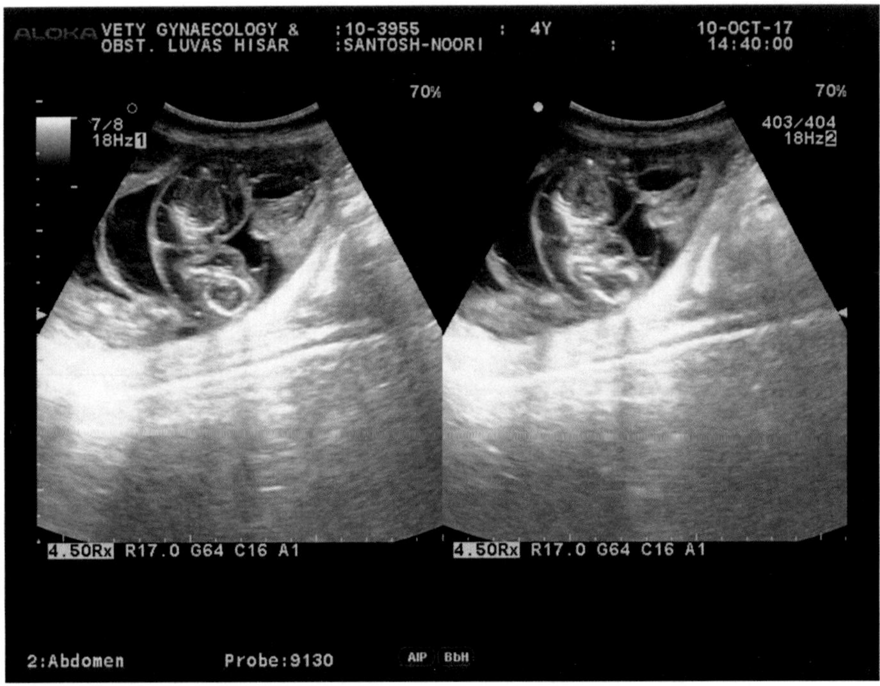

Fig. 42: is showing a case of initation of fetal adsorption in the bitch. Fetus appears transparent indicating digestion of soft tissues.

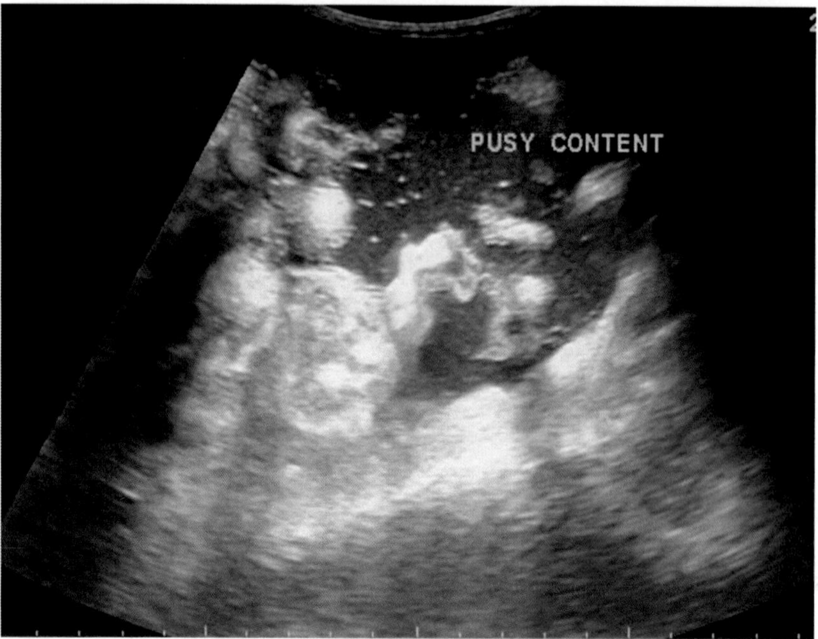

Fig. 43: is showing fetal maeration in a bitch

Dystocia and Fetal Distress

In clinics, many a time cases are presented having problem of dystocia. Owner wishes to know, whether the pups are dead or live in over gestation cases. Sometime one pup is delivered then, it is desired, whether any other pup is inside or not and status of retained pup. In our clinics, such queries were answered through ultrasonography. The diagnosis of status of pups by ultrasonography helped in deciding the line of treatment. Using ultrasound it is easy to detect fetal heartbeats from day 22 onwards and generalized fetal movements from day 28 onwards. In late stage of dystocia, the veterinarian has limited option of focusing either on heart or head at one time. Overlapping of fetal parts make it difficult to ascertain exact number of fetuses in bitch, particularly in advance gestation. Scanning at advanced stage of dystocia case is performed with 3.5 to 5.0 MHz transducers. In such cases, the animal is shaved from lover abdomen and kept in standing or lateral recumbency. In standing position respiratory movements of mother interfere in obtaining good pictures of fetus. Therefore, lateral recumbency is the choice for scanning. If the fetal heartbeat is detected then it should be counted for 5 or 10 seconds and multiply it accordingly to convert it per minutes. From this fetal distress can be easily detected. If the fetal heart rate is less than twice than that of maternal rate, it is an indication of significant distress. In one case of dystocia, the animal showed severe straining without expulsion of pups. On ultrasonographic scanning all the pups found normal with movements and heartbeats, but there were stones in the urinary bladder.

Postpartum Uterine Involution

After a normal whelping, the horns remain enlarged with large amount of fluid inside for a variable time. During first week after delivery, the horns showed multiple layers of varying echogenicity with discrete areas of enlargement with hypo echoic centers at regions of placenta. Measurement of uterine dimensions, particularly of uterine horns is considered helpful in knowing the proper involution of genitalia. Normal size of the uterus, equivalent to that of nonpregnancy i.e. anoestrus, is reached within 4 - 6 weeks of whelping. Postpartum abnormalities

Retained Fetus

Ultrasonographic identification of such a condition is so easy due to the presence of hyper echogenic fetal skeleton. It is cautioned that one should have knowledge of proper identification of fetal skeleton in parts, as full picture of fetus is not possible. Moreover, the machine should have a good quality resolution for that matter. The history of animal should be recorded properly particularly about number of pups already delivered and time of such deliveries along with any abnormal signs after such delivery. As a matter of fact, post-whelping ultrasonography can save the valuable animal and botherations of the owner. During scanning of such animals, sedation of animal should be carefully decided depending upon the nature of animal. However, one should avoid use of anesthetic if possible otherwise.

Retained Placenta

Scanning for such cases can be done using 5.0 MHz transducer. In such cases, there will be echogenicity due to soft tissue debris and blood clots. These will be seen as echoic spots in the unnechoic fluid within uterus. Again, one need to have some experience for such diagnosis otherwise, confirmatory diagnosis of retained placenta will be very difficult.

Visualisation of the Abnormalities of the Non-pregnant Uterus

To obtain images of normal uterus during anoestrus in young nulliparous bitch is difficult due to small transverse diameter, which is less than 1cm, but with good quality machine and transducer of high frequency, it can be visualized on dorsolateral aspect of urinary bladder. In older, pluriparous bitch the scanning of uterus is comparatively easier. The uterine horns and body are having two distinct layers, which in ultrasound can be visualized as centrally, relatively hypo echoic area, surrounded by hyper echoic layer, which are endometrium with serosa.

During oestrus, the transverse section of uterus appears more hypo-echoic as compared to images during anoestrus. Similar images can be obtained, if a bitch is treated with oestrogen. The hypo echoic image is due to oedematous endometrial folds. Further drop in echogenicity of the uterus is seen in mated bitches, during early pregnancy. Images at this time (Few days after mating) show an increase in uterine diameter and presence of anechoic fluid. Continuous scanning during such early phase of pregnancy (also referred as dioestrus phase), there is continuous increase in diameters of cross sections of the uterus, but echogenic appearance of uterine wall remain similar to those images obtained during proestrus.

Cystic Endometrial Hyperplasia (CEH)

This condition is associated with high progesterone concentration and has been observed during luteal phase (diestrus) leading to the cystic hyperplasia of endometrium. Diagnosis of this condition requires skill of the operator and good resolution of the machine. Moreover, in certain cases of CEH changes are not visualized due to small lesions. On careful visualization of the ultrasound image, one may find many small anechoic cysts, which are seen in many sections of the endometrium. In certain cases the dimensions of these cysts reaches up to 5 mm or larger due to coalesce of smaller lesions. There is need to differentiate these cysts from other anechoic images. Infact, these CEH cysts lack a well defined border and are asymmetrical.

Pyometra

Ultrasonograhy is considered as wonderful tool for the diagnosis of pyometra. The ultrasonographic diagnosis of pyometra is possible even before the appearance of the clinical signs. In pyometra, one may obtain transverse or longitudinal sections that show large diameters in comparison to non-pregnant genitalia. Due to increase in diameter of these sections and folding up of the the endometrial layer, images of two or three sections of each horn are seen in a single plane. Enlargement of these sections of uterus depends upon the type of pyometra, being less in open than closed pyometra. In such cases, uterine wall will appear hyper echoic and there will be increase in its thickness. These sections contain anechoic fluid, but mixed with echoic particles of inflammatory debris. In case of pyometra scanning may reveal an outer echogenic area with an inner hypo echoic to anechoic ring and a central hyper echoic zone and they suggested that the hypo echoic region is as a result of vascular enlargement and secretory gland activity. The fluid filled uterus may be mistaken for intestinal loops if there is no marked dilation, but can be differentially diagnosed by lack of peristalsis, which may present in normal intestine. Differential diagnosis becomes easier when distension is of higher magnitude with larger volumes of fluid, which leads to an enhancement of ultrasound image. Differential diagnosis of other causes of uterine fluid accumulation like mucometra and hydrometra is difficult and experience of images of these various conditions.

Vaginal Abnormalities

Vaginal abnormalities like size of vaginal tumors can be assessed e.g. in Trans venereal tumors (TVT) cases.

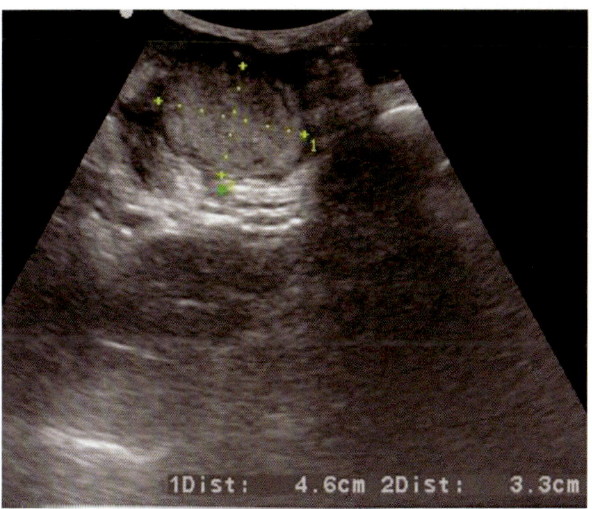

Fig. 44: is showing size of vaginal growth of a TVT case.

13

Ultrasonography of the Liver and the Kidney in Canines

The Liver

The liver is a large and an irregular organ of the body, which extends from the sixth to the eleventh intercostal space. It extends much more caudally on the right side as compared to the left side. Its most caudal part is the caudate lobe whose small portion is visualized encircling the right kidney dorsally at the eleventh intercostal space. The liver of a dog is divided into various lobes namely the left lateral, left medial, quadrate, right medial, right lateral and the caudate lobes which are not identifiable ultrasonographically, except when fluid is present in the abdomen. Otherwise, the position of the different liver lobes can only be approximated. The gallbladder is located to the right of the midline between the quadrate and the right medial lobes of the liver.

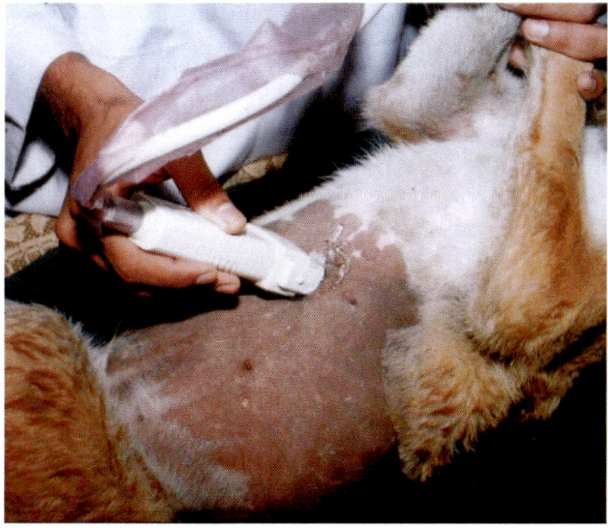

Photograph-13: is showing position of probe for scanning of liver and gall bladder in dog. The probe should be moved from right to left to view all lobes of liver. Right kidney can be taken as landmark and then move from there towards left till spleen and left kidney are seen.

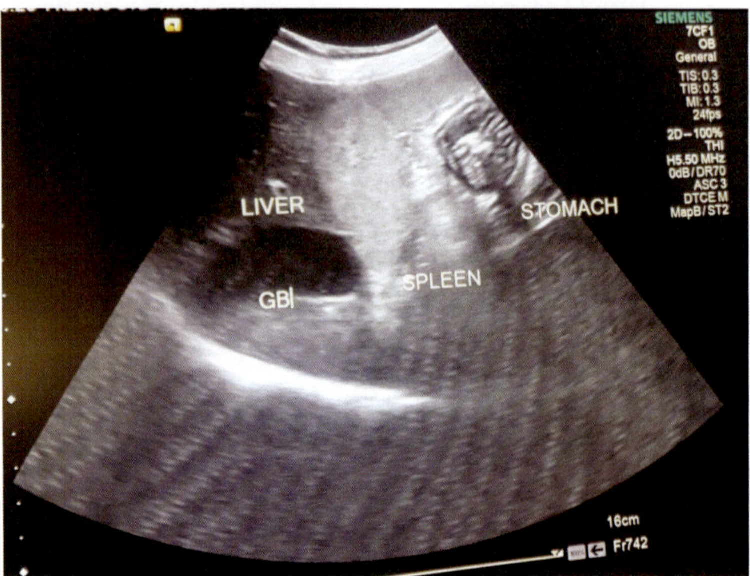

Fig. 45: is showing relative position of Liver, spleen, Gall bladder and Stomach in a dog

Proper preparation of the animal is important to obtain images of abdominal orgnas including that of liver. It is better to keep animal off fed for at east six hours prior to ultrasound scanning. This will avoid interference of gases and solid contents of the gastrointestinal tract. If stomach is taken as landmark of scanning then enough fluid should be given prior to scanning to get good images of stomach. The liver can be easily visualized on one side of stomach. One should be liberal in applying good amount of gel over transducer and body for better image of liver.

It is difficult to get ultrasonic image of the whole liver from one position in a single scan. For better image of liver, it is suggested to place the transducer just behind the xiphisternum in dorsal recumbancy i.e. at ventral midline directing transducer in cranial direction. Middle and Upper portion of liver can be scanned by placing trasducer in between ribs i.e. imaging through the caudal intercostal spaces moving from one to other direction. Suggestive position of dog is dorsal recumbency for keeping transducer at post-xiphoid surface but some ultrasonograpahers suggest lateral recumbancy which is considered as more comfortable position for the dog.

On ultrasonography, the beginner can identify the liver by its location better from position of right kidney or its position in vicinity of stomach as it is identified from its coarse, homogenous echo texture. During scanning of liver, several structures can be visualized in the hepatic parenchyma such as hepatic and portal veins and gallbladder. Identification of hepatic and portal veins can be done by presence of typical echoicity in th wall of the portal vein, which appears as thick, while this is absent in hepatic veins. The walls of hepatic veins are not echoic. The gallbladder can be easily identified and visualized in a scanning. With good quality of machine, hepatic arteries, which are too small might be ualized but difficult to recognize in gray scale 2D ultrasonography. For their visualization the Doppler ultrasonography is better. Parenchyma of a healthy liver is echoic as compared with normal renal cortex and hypo echoic than parenchyma of a normal spleen.

Several abnormalities have been reported in the literature that can be divided into isolated or multi focal and diffused types of abnormalities.

Typical single, focal abnormality is easy to notice as compared to diffused types of abnormalities. Focal abnormalities are generally surrounded by normal hepatic parenchyma. The ultrasonic images might be either hyper echoic, isoechoic or hypo echoic in comparison to the normal hepatic parenchyma. Some times, the abnormality may have several components i.e. increased blood supply, fluid or pus accumulation and or with combination of components. Literature has referred to these type mixed lesions as bull's eye appearance. This type of abnormalities are generally indicative of malignancy, but not always. There might be overlap of the lesions due to a different pathology of liver. Liver cysts are easily identified and are observed during scanning. Liver cysts due to several reasons have been reported in human liver. Ultrasonography might indicate other abnormalities of liver but is not very specific diagnostic for several liver abnormalities. However, apart from cysts in the liver, the ultrasonography can also good for diagnosis of abscess, haematomas, various kinds of nodules including malignant structures. Size of liver cysts might vary and generally appear round with anechoic fluid inside having smooth walls. On other hand, ultrasonography can easily differentiate haematomas from cysts as haemotomas on basis of echogenicities of contents of haematomas. Liver abscesses can be identified easily with their typical appearance, but image of contents of abcess might vary depending upon thickness of pus. Generally, these are irregular and the ultrasonic image might be mixture of hypo echoic and hyperechoic foci and wall may not be well defined. The ultrasound guided sampling or biopsy lead to specific diagnosis.

In diseases that make liver a hyper echoic in general, the echoic appearance of wall of portal vein are not propely identified.

Fatty liver can be easily identified. On the other hand, other conditions such as liver fibrosis and lymphosarcoma which make liver hyhperechoic and decrease size of liver with irregular margins.

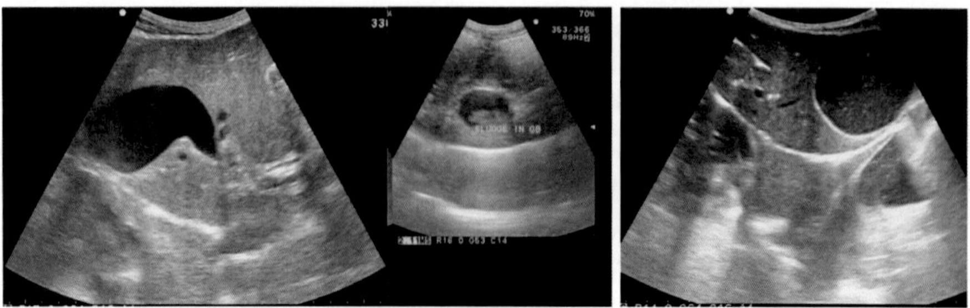

Fig. 46 & 47: are showing fully dilated gall bladeerr with fatty liver (left) and sludge in Gall bladder (middle and right) in a dog

In ascites, the cirrhosis of liver can be seen as hyperechoic decreased size of liver with separate liver lobes. The other diseases of liver reported in literature are end stage liver and visualization of many portal veins i.e. starry sky appearance of liver.

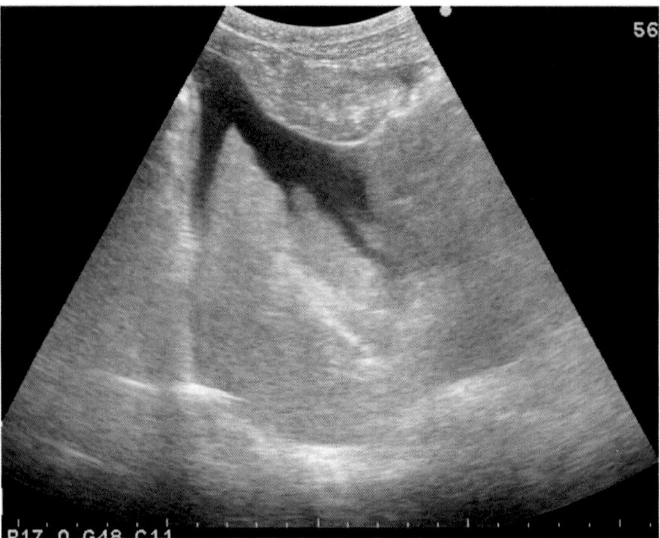

Fig. 48: is showing liver lobe in a case of ascites in a dog

Hepatitis is common contion of liver, howerver, wrong setting of machines particularly very high gain can show liver as like hepatitis. Therefore, setting of machine should be same for all diagnostic purposes.

The gallbladder, which can be very easily visualized as anechoic structure towards the right of the midline. In cholecystitis, the walls of gall bladder appear hyperechoic. One should be careful to diagnose it as hyperechoic lower wall (wall under bile) appear hyperechoic. It is an artifact and not abnormalities.

Sludge has been commonly seen in gall bladder of dog, however, gall stones were not observed in domgs in our clinics. In human, biliary duct dilation has been reported in chronic obstruction. A thickened gallbladder wall due to edema is seen in cholecystitis.

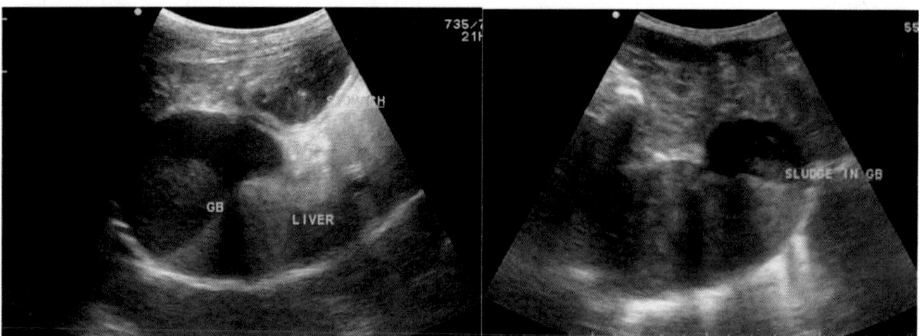

Fig. 49 & 50: are Showing sludge in gall bladder of a dog

Double liver can be observed as an artifact. This occurs if transducer is curved too much from area caudal to xiphisternum. One liver is seen one both side of diapharam, which is not normal. It is scanning artificat.

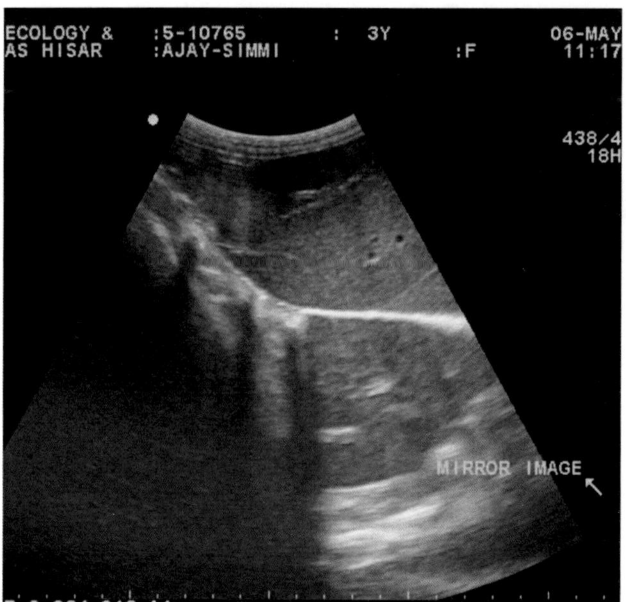

Fig. 51: is showing mirror image of liver across diaphragm as an artifact in a dog

The Kidneys

The ultrasonography of kidney requires understanding of location and architecture of kidneys. The left kidney is located some what caudally than the right kidney. The right kidney is somewhat bisected by 13th rib. Therefore, for taking images of right kidney, one needs to curve the transducer as the cranial pole of the right kidney is far and positioned in the renal fossa of the caudate lobe of the liver. The positioning of transducer and its proper angle is required to take full image of kidney. It is easy to obtaine full scanning of the left kidney that is located below the second to fifth lumbar vertebrae.

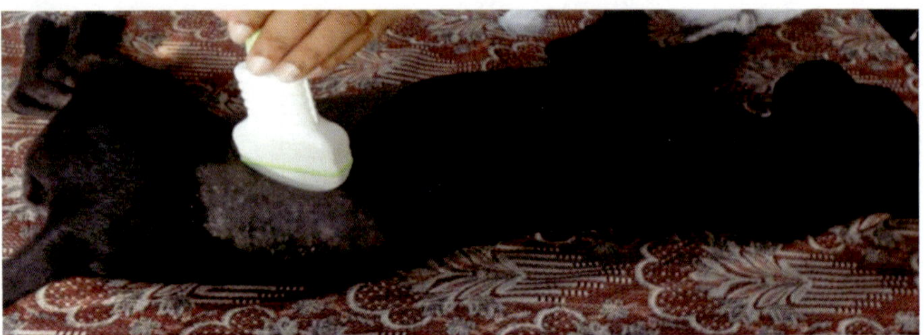

Photograph-14: is showing position of transducer for imaging lefft Kidney in lateral recumbency. The transducer is placed below the transverse process of first to third lumbar vertebra

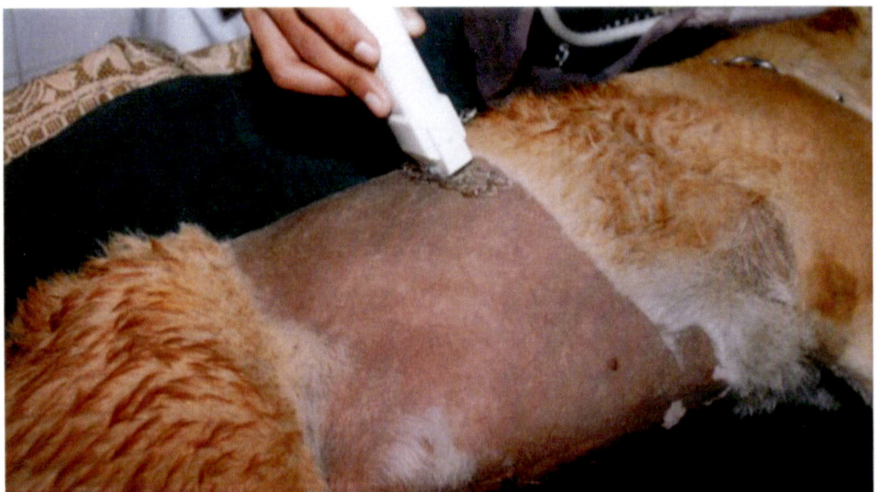

Photograph- 15: is showing position of transducer for scanning right kidney. The transduce should be curved from this position.

Simply, for ultrasonographic image of right kidney, one should put transducer caudal to last rib i.e. caudal to the last intercostal arch. The imaging left kidney, transducer should be placed below the lumbar vertebrae without angling the transducer. Intercostal approach has also been recommended for right kidney, but for this one need small transducer that fits into intercostal space. Sometime, imaging of right kidney becomes difficult, especially, when there is gas in the colon/abdomen.

During each scanning, the kind of images obtained might vary as kind of image depends upon the region focussed and beam angle. Images in three planes are obtained i.e. longitudinal, transverse and dorsal planes. Changing angle of transducer, the planes can be further divided into oblique planes. The angle of image of kidney may change diameter of different regions. Therefore, proper positioning and focusing is important to obtain proper dimensions of kidney. One should be knowledgeable about internal anatomy of kidney. In kidneys of dog, the cortex is on the periphery while, medulla is located after cortex towards center and the sinus is in the centre or eccentric position. Therefore, from one end to other end of image, it is cortex layer, then comes medulla, then sinus in the center, again other side of medulla and then cortex.

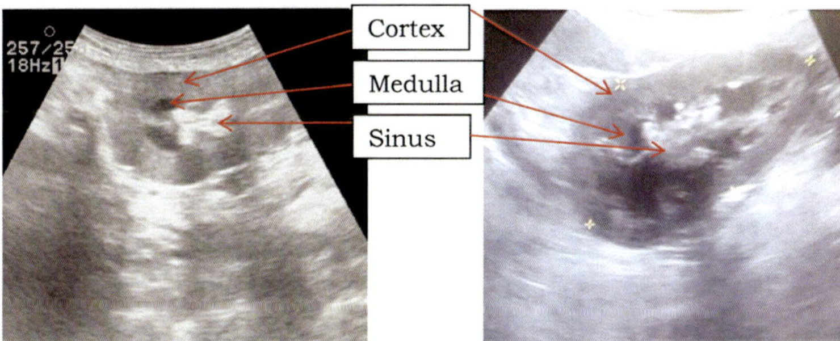

Fig. 52 & 53: are sowing internal anatomy of kidney in a dog

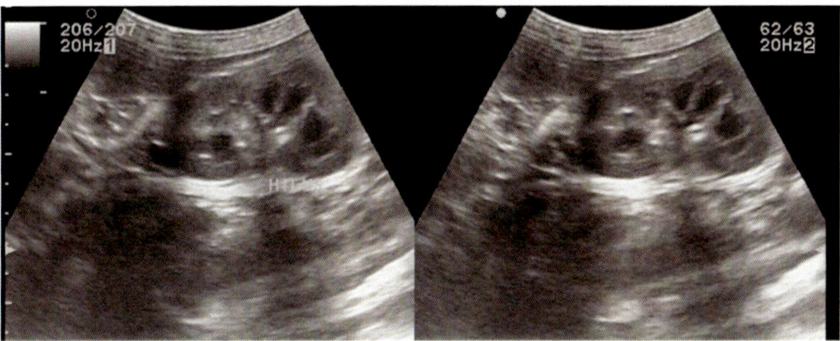

Fig. 54: showing pyramids of left kidney with spleen portion on top left in a dog

Echoically, the cortex is hyper echoic, medulla is anechoic. There is a well-defined corticomedullary junction. Renal sinus appears as a hyper echoic area due to deposition of fat in this region and it forms a c-shape in the transverse scan of the kidney. The medulla shows different compartments which are infact, the linear echoes formed by the pelvic diverticulae and interlobar vessels. Diagnosis of abnormalities of kidneys:

There are different diseases that increase echoicity of cortex of kidney showing increased cortical echoicity with enhanced corticomedullary distinction. Some of common abnormalities are nephroliths, nephritis, hydronephrosis, end stage kidneys, presence of leptosipra in the kidney etc.

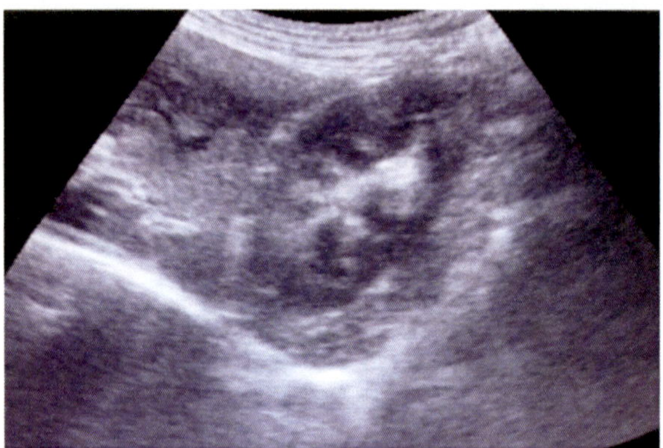

Fig. 55: is showing nepthritis in a dog, where there is increase in echoicity of renal parenchyma and less differentiation of cortex and medulla.

In hydronephrosis, there is loss or dilation of compartments of medulla. In several cases, dilation of the renal pelvis can be obseved as an anechoic separation of the central diverticular reflections. On other hand, pelvic dilation is easily seen central anechoic region within the kidney in hydronephrotic kidney.

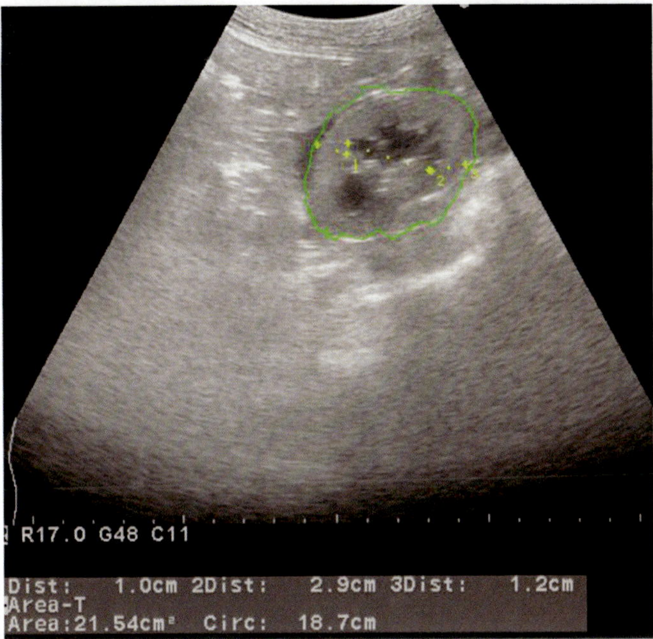

Fig. 56: is showing pelvic dilation in left kidney in ascitis case of dog.

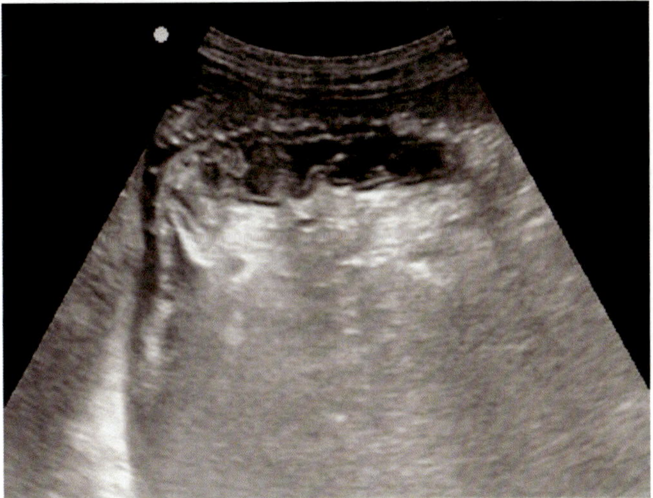

Fig. 57: is showing end stage kidney in a dog. In end stage kidney, there is complete loss of architechture of kidney.

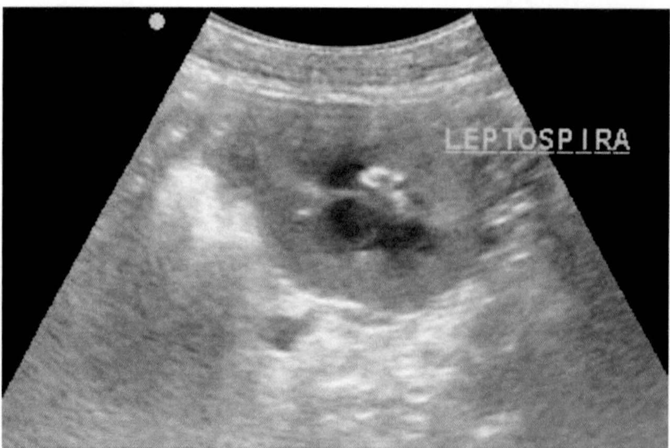

Fig. 58: is showing presence of Leptospira in a kidney of a dog

Different kinds of renal cysts have also been reported that might be single or multiple, large or small and unilateral or bilateral. Cysts usually are round, anechoic, have smooth thin walls and show distal acoustic enhancement.

The nephroliths can be seen in the anechoic part of medulla or some time partially embedded in the cortex. The liths can be single or multiple. To get full size of liths, one has to focus from different angles, but 3D scanning is better for this purpose.

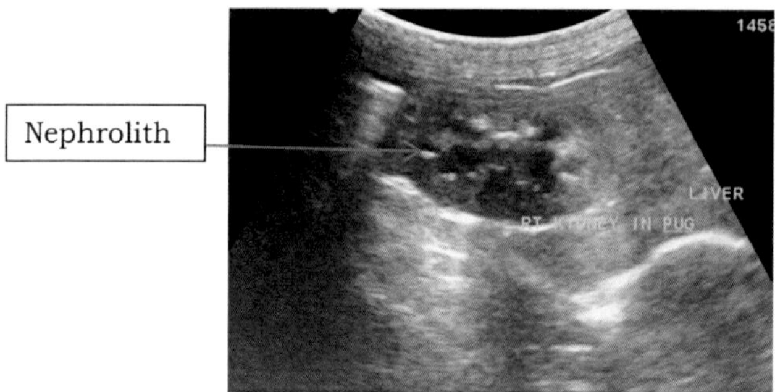

Fig. 59: is showing multiple liths in a kidney of a young age dog

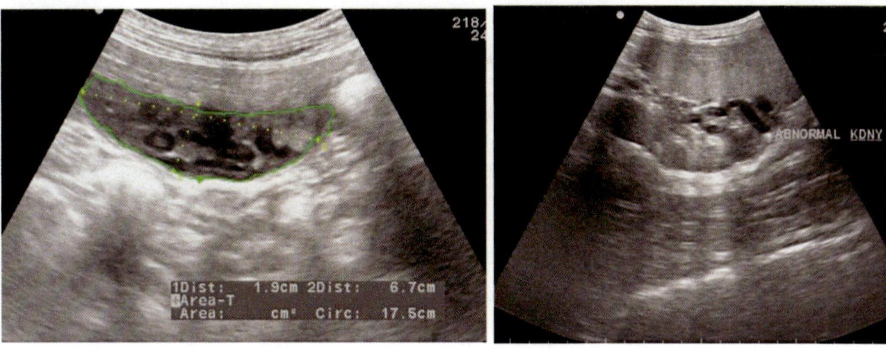

Fig. 60 & 61: are showing abnormal kidneys having cystic appearance in a dog.

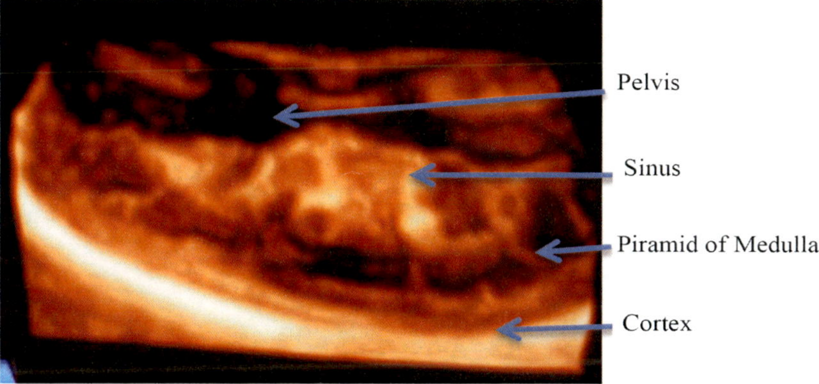

Fig. 62: is showing colored 3D architechture of ultrasonic section of kidney in a dog

14

Ultrasonography of Stomach in Dogs

Ultrasonographic images of stomach are required to diagnose ingestion of any abanormal materials by dog. When content of stomach are dry or there is gas in stomach then, it is difficult to get proper image of stomach. Offering water to animal or oral administration of water with the help of syring is also helpful in getting the images of content of stomach. Some time, while doing ultrasonography, movement of contents of stomach can be visualized. Stomach can be images between liver and spleen. The size depends upon contents or intake of food by dog. For proper image of gastrointestinal tract including stomach, there should not be gas in GIT. One should remember that ultrasound beam does not travel in gas. The best medium of ultrasound beam is liquid.

Scanning Technique

If possible, fasting for 12 hours is good to diminish the amount of gastrointestinal gas. However, non-fasted dogs, when offered water may show adequate image quality. The water may be administered in quantity of 15 ml per kg of bodyweight. Removing the stomach gas using an orogastric tube has been recommended tto enhance the visualization of lesions of the stomach, however, this is not in practice in India.

Fox X-ray, Barium sulphate is useful, but it is undesirable for ultrasonography Scanning position of the dog depends upon operator preferences, dorsal recombancy is alright, but position may be changed to get image of other areas of interest. Alternate positions that can be used might be either right lateral recumbency for assessment of pyloric area or left lateral recumbency for the fundus, and standing position for the body. Standing position has its own advantages and disadvantages.

Position of the transducer should be caudal to the costal arch in a dorsal recumbancy. The gastric area can be examined from right to left, angling the beam craniodorsally. If the stomach contents are not too dense, then the rugal folds can be seen. If the stomach contents are liquid, then regular contractions can also be seen on screen.

A 5.0 – 7.5 MHz transducer is advisable for ultrasonographic examination of the GIT, but a high resolution 7.5 MHz transducer is required to identify the five ultrasonographic layers of the stomach.

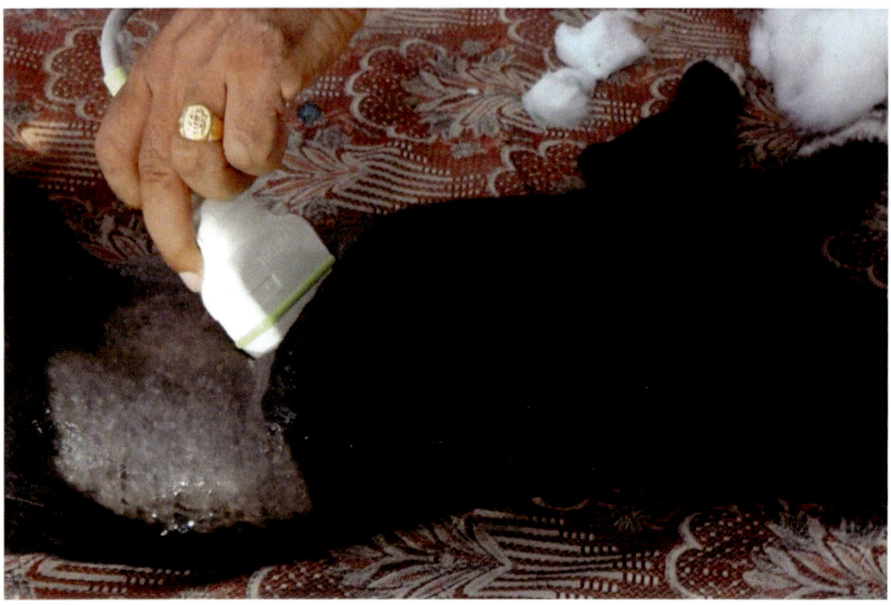

Photograph-16: is showing left side position of probe for stomach scanning. Movement of probe will be required towards upper side and middle side depending upon fullness of stomach.

Details of Observations

The five layers of the stomach have alternating echogenicity. The innermost hyper echoic layer corresponds to the surface of the mucosa; the innermost hypo echoic layer represents the mucosa; the middle hyper echoic layer is the submucosa; the outer hypo echoic layer is the muscularis propria; and the outer hyper echoic layer is the subserosa/ serosa. Histological sections of the different segments have shown same layer distribution.

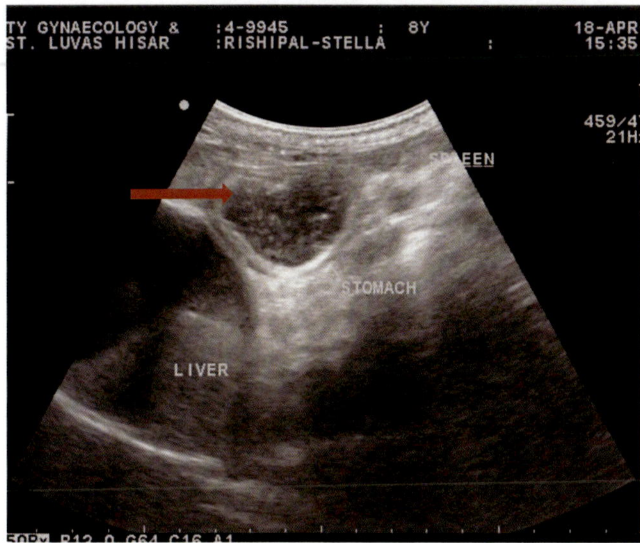

Fig. 63: is showing stomach of dog (arrow) with its contents

The rugal folds may be seen as a star shape in cross-section with a central fluid or gas echogenicity. Fluid or gas may be seen swirling around in the stomach as a result of peristalsis.

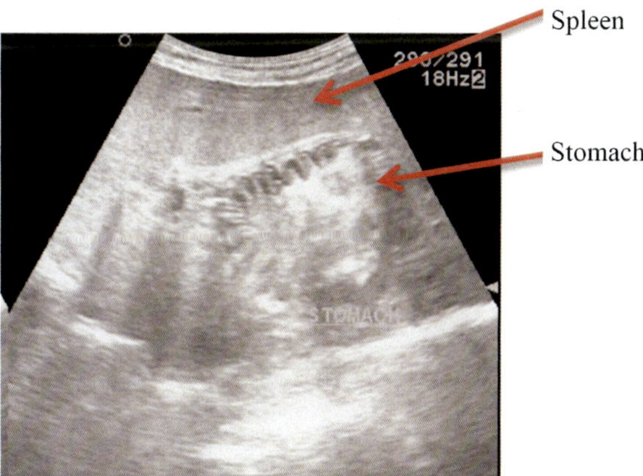

Fig. 64: is showing cross section of stomach with rugal folds

The peristalsis can be routinely observed in the stomach. The mean number of peristaltic contractions of the stomach and proximal duodenum are four to five contractions per minute. The normal stomach ranges from 3 mm to 5 mm in thickness.

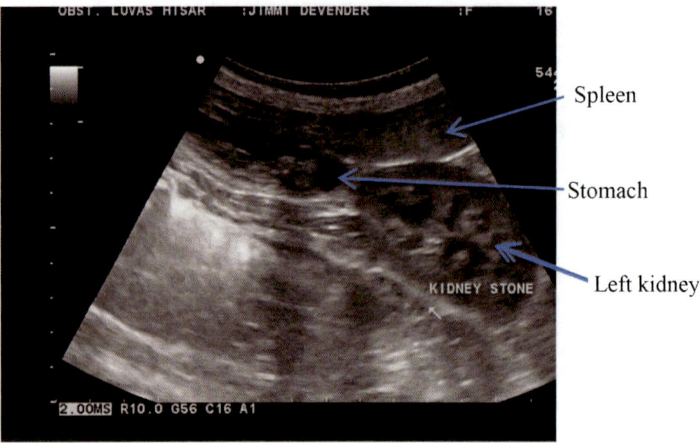

Fig. 65: is showing relative position of stomach, spleen and left kidney in a dog

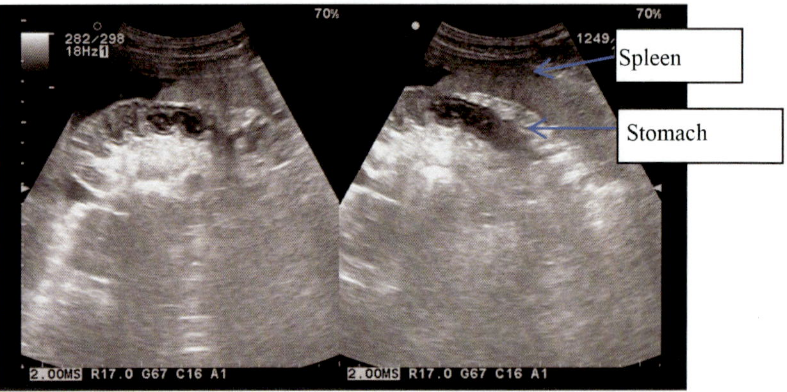

Fig. 66: is showing stomach with different portion of spleen in a dog

Abnormality of Stomach

Luminal contents can be assessed. Ultrasonography has been found to confirm the presence of foreign material within the stomach when the radiographic findings are not clear. There was a case of ingestion of mango kernel that was not detected by X-ray, but was diagnosed with ultrasonography. In some dogs, presence of metallic objects or pieces of solid concrete have been recorded on our facility. For detection of lesions in the wall, one needs transducer of higher frequency that could be 7.5 MHz or more. Detection of wall lesions (abrasions) have also been seen in gastritis cases, where vomition was the history. The acoustic patterns arising from each foreign body depends upon its nature i.e. kind of material. Some pup ingest small plastic balls, while playing. This may not be picked by radiography, but can be identified with ultrasonography.

The ultrasonographic appearance of diffuse inflammatory disease of the stomach depends on the cause and severity of the process. Inflammatory processes may produce a gastric hypo motility, with or without thickening of the wall. Other signs that may be observed in inflammatory disease are an altered echogenicity of the layers, such as a hyper echogenicity of the internal aspect of the mucosa and poor wall layer definition. Associated mesenteric lymphadenopathy may also be identified.

Ultrasonography has been reported to be more sensitive than survey radiography for the identification of gastric lesions associated with uremia in dogs. A thickened gastric wall, thickened rugal folds and a hyper echoic line at the mucosal-luminal interface due to mineralization of the gastric mucosa have been reported in severe uremia caused by chronic renal disease. A layered mineralization of the gastric mucosa should be considered an unfavorable sign in uremic dogs.

Ultrasonography can provide morphological assessment of gastric neoplasms even without any specific patient preparation. This may prompt a tentative diagnosis of gastric neoplasia and thus stimulate further investigation. In gastric tumors; the echogenicity of the wall lesions is variable and independent of the type of tumor. But in majority of the dogs suffering from gastric neoplasia, sessile masses involving all layers of the gastric wall may be imaged. Along with, ulceration and lymphadenopathy could be observed. But lymphadenopathy may not be identified sonographically in leiomyoma and leiomyosarcoma. Signs of extension of the lesion through the serosal surface of the stomach has been identified sonographically in dogs with carcinoma.

Ultrasonography is a recommended screening procedure for cases where gastric neoplasia is suspected. Moreover, the likely findings of gastric neoplasia i.e thickening of the gastric wall, loss of gastric wall layering and enlargement of regional lymph nodes can be assessed better ultrasonographically than on survey radiography. Ultrasonography can assist in obtaining diagnostic samples and in clinical staging of the tumor.

Even when gastric disease is suspected the entire abdomen should be scanned. Hepatic, gall bladder and pancreatic disease can produce GI signs and are easily detected. Lymph nodes can be detected and is evidence of systemic disease. Ultrasound guided fine needle aspiration biopsy and core biopsy provides several potential advantages as an initial diagnostic tool for infiltrative diseases of the stomach. It is safe, rapid and achieves a high percentage of correct diagnosis. It may allow for planning of definitive surgical treatment or radiation therapy. This approach can provide a prognosis for the animal prior to embarking on a more costly or risky procedure. Ultrasound guided procedures can assess areas unreachable with an endoscope, avoid perioperative morbidity in case of lymphoma, and minimize or avoid the use of general anesthesia. Ultrasonography is a useful tool for the detection and diagnosis of various disorders of the stomach. It is not meant to replace diagnostic radiology but to complement it. Ultrasound is operator dependent. The quality of the image and the information gained are directly related to the ability of the person doing the study.

15

Ultrasonography of Testis and Accessory Sex Glands in Bull

Ultrasonography in bull is helpful in diagnosis of problems of testis and accessory glands. This can also be used in assessment of development of these organs. Such developmental assessment, when combined with endocrine data, is useful to know age of puberty, sexual development and dimension of these organs. The ultrasonographic images and data are available for testis, bulbourethral gland, prostate and seminal vesicles in the bull. Ultrasonography of testis of bull can be useful to assess developmental changes, testicular dimensions and pathological conditions of the testis viz. tumors, lesions, abscess etc. Epididymis can also be examined for change in lumen, inflammation, granuloma etc. The accessory sex glands can be better viewed in ultrasonographic scanning to assess their dimensions, normal developmental changes and alterations in their morphology. Before, ultrasonography is undertaken, one should thoroughly revise the knowledge of sectional anatomy and location of these organs.

Testis

Location: The testis, in the bull, is located in the scrotum. In abnormal conditions, the testis might be retained in the inguinal canal or abdomen. Before, ultrasonography, location of the testis should be assured. For this, the testis should be palpated to assess their pliability and rule out adhesions with the scrotal wall. In normal condition, the testis should be freely movable in the scrotum, soft of palpation and both the testis should touch the bottom end of scrotum.

Morphology: The testis is covered wit tunica vaginalis layer, which is the extension of peritoneum. If we take out the testis and scan it, followings are layers from outside to inside. Tunica albuginea testis and parenchyma. Parenchyma is divided into segments, which contains seminiferous tubules. The central tube, rete-testis, collects sperms from seminiferous tubules.

In ultrasonographic scanning, the normal parenchyma appears as homogenous, while the rete-testis appears as bright, echoic central line-cavity.

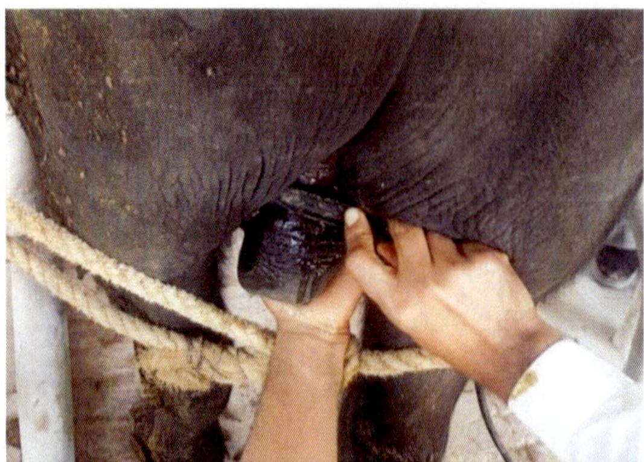

Photograph- 17: is showing position of probe for horizontal scanning of testis.

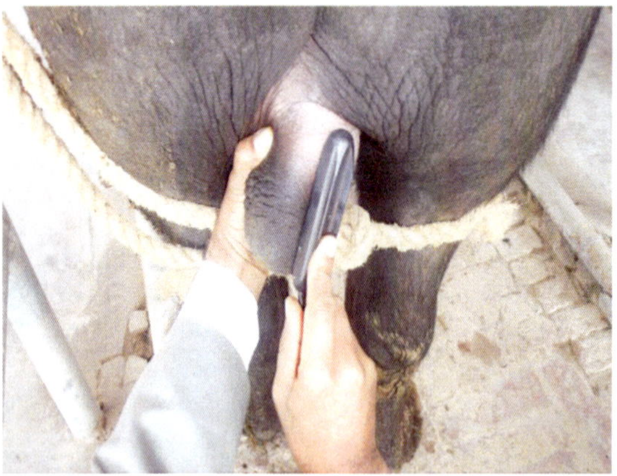

Photograph-18: is showing scanning of testis in longitudinal position in a buffalo bull

Procedure to Scan Testis

The bull should be properly restrained in animal chute/cattle crush.

Generally, there is no need of shaving the surface of scrotum, however it should be ensured that there is no hairs on scrotal surface.

The temperament of animal should be tested by touching the back-rump portion, thigh, and scrotum. The animal should be petted and touched at many places, particularly hind limbs to make the animal aware that you are there at the back portion of the animal. If animal allow you to touch scrotum and hind limbs without reaction, that means, animal is co-operative, otherwise light sedation may be used. Do not use high epidural anesthesia.

Apply ultrascan-gel over the surface of scrotum and active face of the transducer. The testis should be scanned in two planes viz. longitudinal (lengthwise) and vertical (widthwise).

In most of the cases, 5.0 MHz frequency transducer is suitable, however for detailed study of the organ 7.5 MHz transducer should be used.

Pull the scrotum from the back.

Push one testis up, but keep other one in the bottom of the scrotum.

The testis should be scanned at three places. One central scanning to get rete-testis in the picture, while two scanning on either side of it.

The scanned images should be stores on videocassettes or compact discs or floppy discs. The pictures can be printed immediately, however the printer should be set at its optimum level. In our hand, the photography done from the screen of monitor through playing of storage disk provided better photographic depiction of organs.

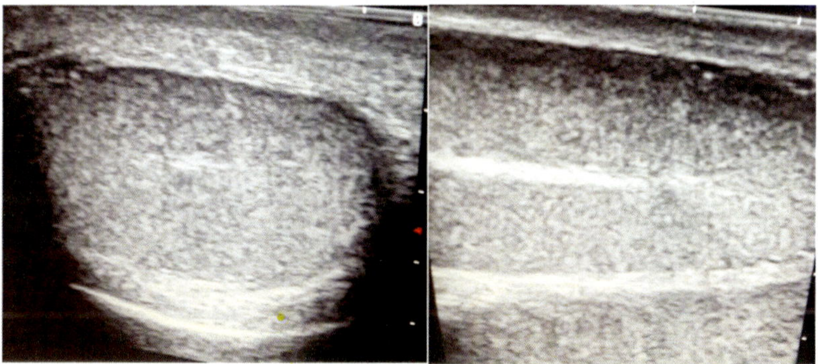

Fig. 67 & 68: are showing cross secion **(left)** and longitudinal section **(right)** of testis of buffalo bull.

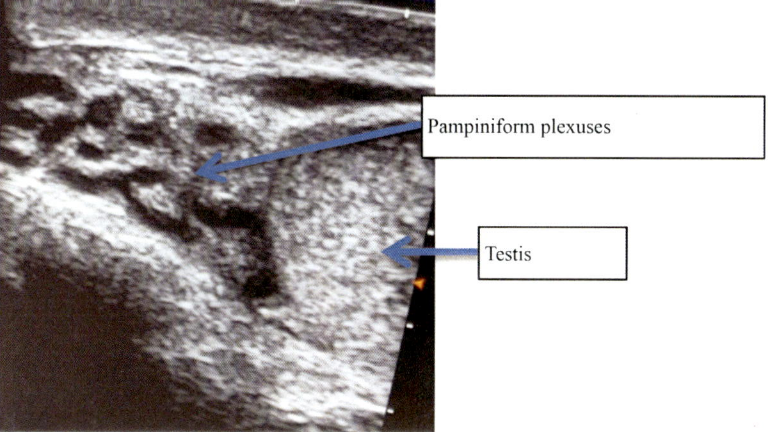

Fig. 69: is showing cross section image of Pampiniform plexuses on top to testis in a buffalo bull

Accessory sex glands

Before ultrasonography of accessory sex glands (Bulbourethral gland, Prostate and Vesicular gland), rectal examination of the bull should be done to palpate these glands. Following is the rectal procedure of bull for the purpose.

Wear disposable, full sleeve gloves and lubricate with liquid paraffin.

Pass hand into the rectum and try to palpate the bulbourethral (Cowper's) glands. These glands are paired glands and located near the ischial arch, therefore, try to palpate them just close to (near) the entrance of rectum on the floor of pelvis. These glands in bovines are covered with thick muscles; therefore, these glands can be palpated as round hard mass and not individual one. Theoretically, these glands are ovoid in shape in bull.

Urethra can be palpated as cylindrical, tubular organ at the floor of pelvis. To identify urethra, massage on its dorsal surface, the urethra starts pulsating.

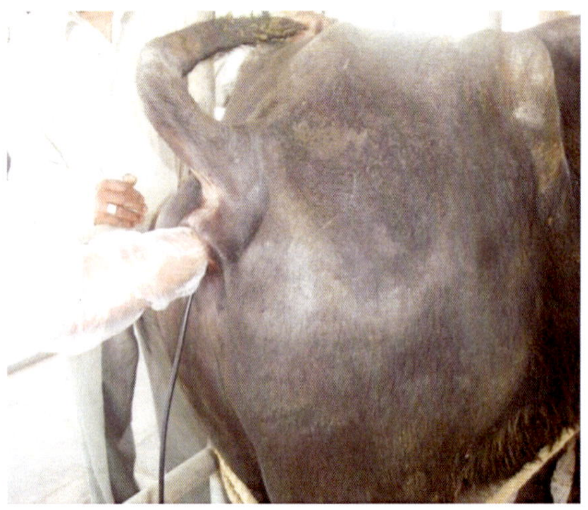

Photograph-19: is showing ultrasound scan position for accessory sex glands beginning very close to entry for bulbourethral gland followed by prostate, then seminal vesicles on the side of urinary bladder.

The prostate in bull can be palpated caudal to urinary bladder. It is diffused around urethra; so, it will be felt a thick band around the urethra.

The seminal vesicles, which are termed as vesicular glands in bull can be palpated on pelvic floor, lateral to the prostate gland. Try to spread your fingers from the prostate towards lateral direction. The vesicular glands can be palpated as round and lobulated organs.

Procedure for ultrasonography

Select a good quality transducer of switch-able frequencies between 5.0 and 7.5 MHz. After restraining the animal in chute, fecal material should be evacuated from the rectum, but avoid ballooning of the rectum.

Apply gel on the active surface of the transducer and hold it in a cup shaped hand. Insertion of some quantity of gel into rectum may be considered, but most of the time only gel application on the transducer is good enough for the procedure.

The knowledge of location and morphology of the accessory glands is must; that can be advantageous to get good quality images of these glands.

The transducer is moved slowly along the rectal mucosa. The bulbo-urethral glands appear round mass, close to the ischial arch. As the transducer is moved cranially, the longitudinal image of urethra is obtained. Move the transducer further to get cylindrical image of prostate gland. Alternatively, one can get image of urinary bladder and then comes back to neck side to get image of the prostate gland.

Prostate glands of buffalo bull

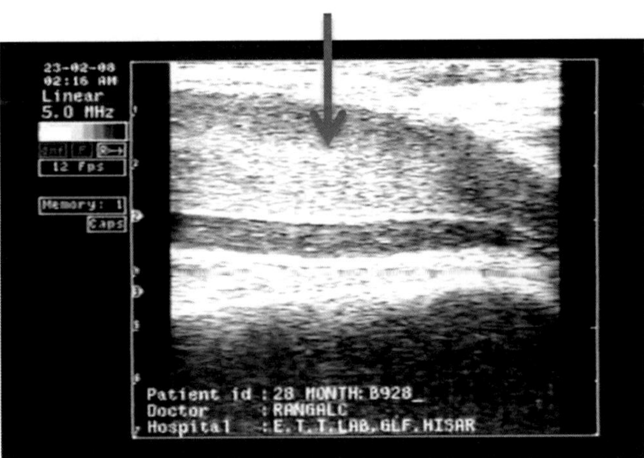

Fig. 70: is showing image of prostate surrounding pelvic urethra.

To obtain images of vesicular gland (seminal vesicles) one has to move in the lateral direction from Prostate gland. The vesicular glands appear lobulated in adult males, while it may be round in immature growing male-calves.

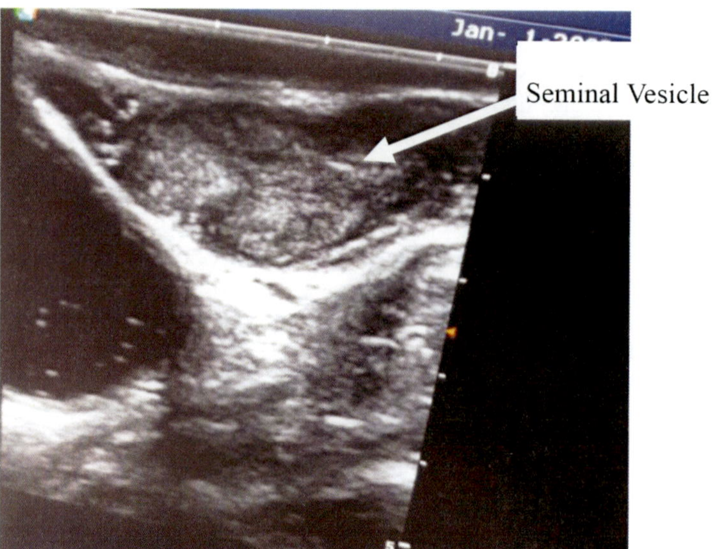

Fig. 71: is showing lobulated image of seminal vesicle on lateral side of urinary bladder in a buffalo bull

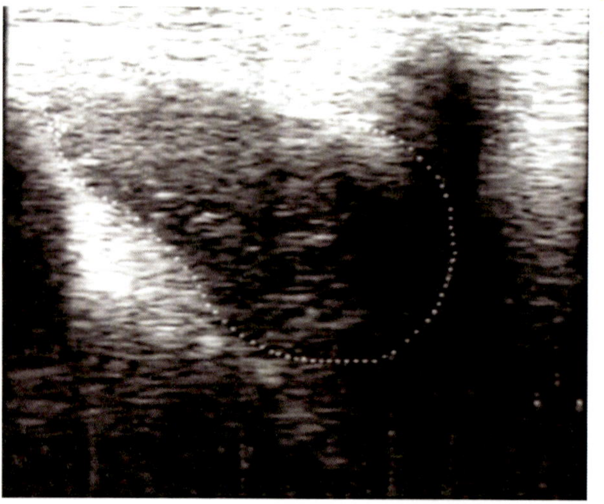

Fig. 72: is showing image of bulbourethral gland in a buffalo bull

16

Use of Ultrasonography in Equines to Assess Reproductive Organs

Real time ultrasonography provides means to directly image the reproductive tract. An extensive amount of information can be gained from even a single examination, including determination of ovarian (luteal or follicle detection) and uterine status, detection and monitoring of ovarian and uterine pathology, monitoring the post-partum mare, detection of twins, prediction of ovulation, fetal growth and development and determination of sex of fetus. When first introduced the use of ultrasonography was primarily limited to early pregnancy diagnosis only, with advancement of knowledge and interpretive skills, the diagnostic ultrasonography has a much more application in reproduction. Some of the applications are discussed in this lecture.

Procedure

The procedure and precautions for intrarectal ultrasonographic examination are similar to those for rectal palpation, and no additional restraint is required. The transducer should be protected by the examiner's hand to prevent trauma to the rectal wall, and the transducer should be well-lubricated. Care should be taken to avoid fecal material attaching to the transducer. After evacuating fecal material from the rectum, the probe is introduced and moved across the reproductive tract in the following pattern: uterine body, right uterine horn, right ovary, right uterine horn, uterine body, left uterine horn, left ovary, left uterine horn, uterine body then cervix. Good contact must exist between the transducer and rectal wall. Air in the rectum or a gas or fluid-filled loop of bowel will result in a distorted image. To minimize scanning errors, principally those of omission, it is recommended to conduct the same scanning procedure during each examination.

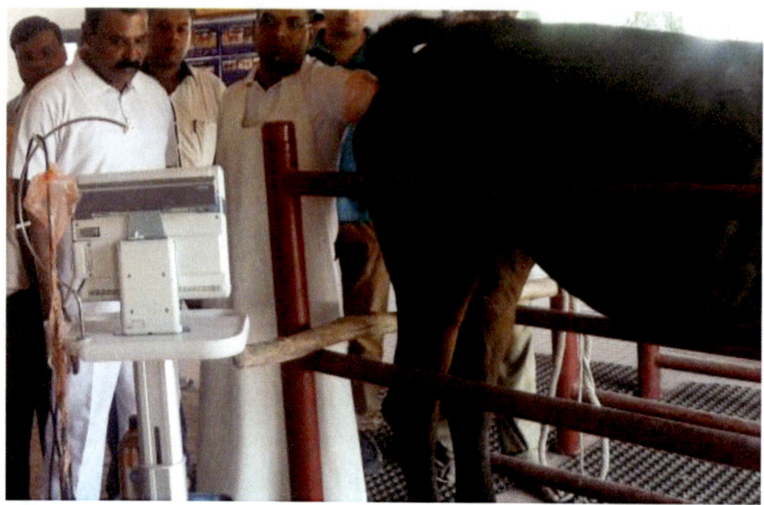

Photograph- 20: is showing position of ultrasound machine and restraining of mare for transrectal ultrasonography.

Equipment

The two major types of real-time ultrasonographic transducer used for reproductive examination of the mare are linear and sectorial. Physical arrangement of the crystals within the transducer determines the pattern by which sound waves are propagated from the transducer. In linear-array scanners, the width of the rectangular ultrasonic beam corresponds to length of the active or crystallized portion of the transducer. A linear-array transducer is oriented in the longitudinal plane with respect to the mare's body. Therefore, images of the cervix and uterine body are longitudinally – oriented and those of the uterine horn are cross-sectional. Images of tissues closest to the transducer are at the top of the screen. Sector scanner produces a beam that is triangular in shape, because sound waves radiate from a single point or source. The sound beam generally travels transversely to the mare's body and consequently, images of the cervix and uterine body are cross –sectional, while images of the horns are longitudinal or oblique.

Resolution, which is the ability to detect small differences in tissue density, depends on frequency of sound waves. Higher frequency provides greater detail and lower frequency provides greater tissue penetration. Ultrasonographic frequencies are measured in megahertz (MHz, one hertz (Hz) = one sound wave/second). The lower frequency transducer (3 and 3.5 MHz) is suited for viewing larger structures at a greater distance from the transducer than the 5 or

7.5 MHz Transducer

Higher frequency transducers (5 or 7.5 MHz) are most useful for detailed study of structures close to the transducer. Five MHz transducers can be used to detect a conceptus on day 10 (McKinnon *et al.* 1988), follicles as small as 3mm (0.12 inches) and presence of a CL throughout most of diestrus (McKinnon *et al.* 1987). In comparison, a 3 or 3.5 MHz transducer can be used to detect a conceptus at about day 13 to 15, follicles approximately 6 to 8 mm (0.24 to 0.32 inches) in diameter, and the presence of a CL for 5 to 6 days post –ovulation. The principal uses of lower frequency transducers are to study an older fetus by either intrarectal or abdominal scanning or pathological conditions elsewhere in the body, such as liver abscesses.

Folliculogenesis

Follicular dynamics is important to study different events of an oestrus cycle. The basic aim of ultrasonography of follicle is to breed the mare within a short time before ovulation. Approximately 7 days before ovulation (Day of ovulation + Day 0) dominant oestrus follicles has a mean diameter of about 25 mm. They then grow at 2 to 2.5 mm per day and reach their maximum diameter of 41 to 45 mm at 24 to 48 hrs before ovulation (Pierson and Ginther, 1985; Pal and Gupta, 2004). In most cases no further growth occurs during the last 1 to 2 days before ovulation (Palmer and Driancourt, 1980). The diameter of the preovulatory follicle often remains static. Only occasionally follicles are observed to be smaller at the time of ovulation.

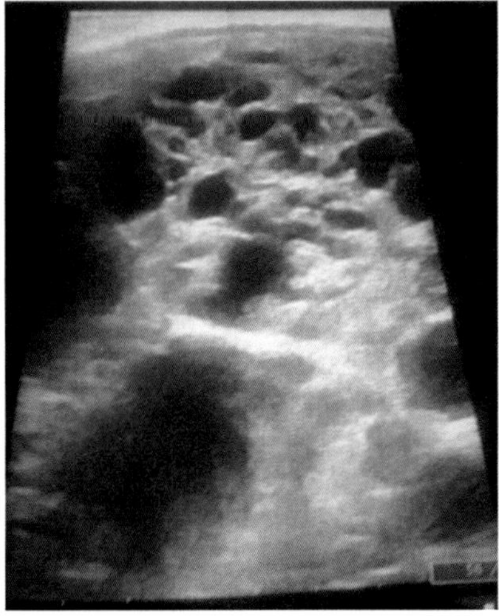

Fig. 73: is showing multiple follcles of various sizes in an infertile mare suspected for granulosa cell tumor.

The upper limit of the normal ovulatory follicles lies between 55 and 58 mm. At the time of ovulation of a single follicle it usually had a diameter of 40 mm or more. When double ovulation occur the diameter of the preovulatory follicles may be smaller (between 35 and 40 mm) than those of single follicles. The majority of dominant follicles are distinctly round about 3 days before ovulation. During the days until ovulation the estrous follicle will change to a more oval or irregular shape. On the day of ovulation only about one third of the oestrous follicles will be round in shape. Sirois *et al.* (1989) indicated that mares exhibited either one (71%) or two (21%) follicular waves per cycle.

Ovulation

The disappearance of previously detected preovulatory-sized follicle is an indication of recent ovulation. The actual collapsing of the ovulating follicle can only be demonstrated by chance if the mare is examined very frequently. It usually only takes seconds to minutes. The subsequent formation and identification of a corpus luteum confirms ovulation. During the course of ovulation, two patterns of follicular evacuation had been reported (1) an abrupt loss of follicular fluid in which majority of fluid (>90%) disappeared within one minute and (II) a slow, gradual loss of follicular fluid in which 80% of the fluid disappeared within four minutes. Dowsett *et al.* (1994) recorded that 60% of ovulation occurred in the last two days of oestrus and average size of follicle on the day of ovulation was 41.5 mm (range 31 – 51 mm).

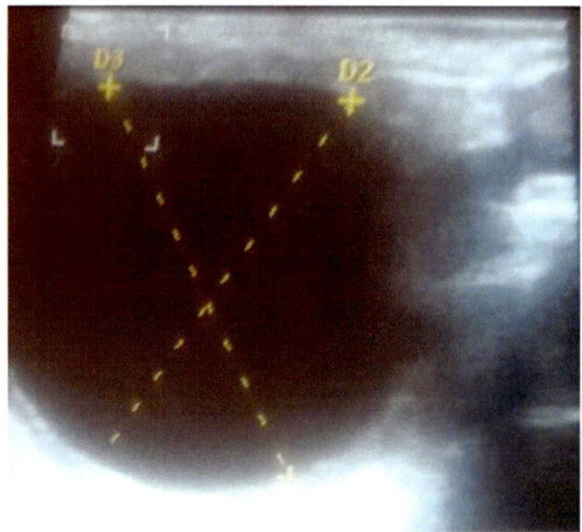

Fig. 74: is showing full size follicle of 4.83 cm (measurements not shown) in a mare.

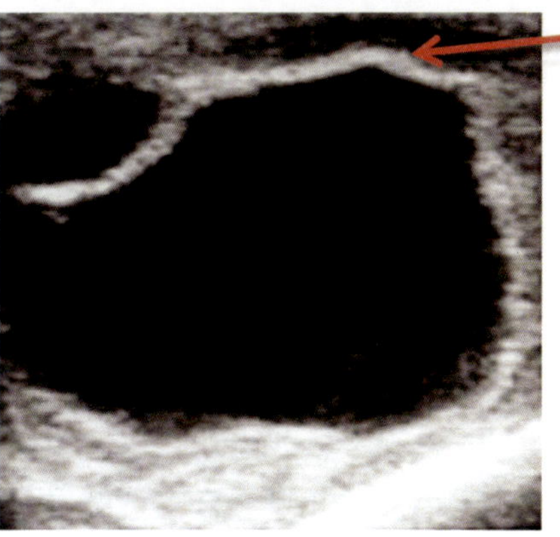

Fig. 75: is showing ovulatory follicle (See collapsing walls)

Predicting Ovulation

Since female equids are sexually receptive for several days and it is difficult to decide when to inseminate a mare. Mating/insemination during first few days of oestrus leads to low fertility. The ability to detect the time of ovulation has an important practical utility in equine reproduction as ovulation takes place before the end of oestrus. Prior to the introduction of efficient ultrasound scanning techniques, palpation per rectum was the traditional method by which ovulation was detected (Hughes *et al.*, 1972). Now the ovulation is predicted on the basis of rectal palpation and scanning.

It has observed that a combination of softening of a large follicle and/or pain at follicular size (as determined by palpation per rectum) and a marked change in the shape of follicle (equal to or greater than 40 mm) and, increased echogenecity of follicular fluid could be useful in predicting ovulation with in a 24 hr period for most of the mares.

Multiple ovulation: Female equids generally ovulate one follicle during an oestrus cycle. Thoroughbred and draft mares have the higher incidence of multiple ovulations. (Ginther, 1992). Double ovulation may be synchronous (occurring on the same day) and asynchronous. It is difficult to detect the multiple follicles by rectal palpation, especially when they are on the same ovary. But, using ultrasonography multiple ovulations can be detected very easily and such females are not inseminated/mated.

Time of Insemination

In cattle and buffalo, insemination is done 12 hr after the end of oestrus and it is thumb rule. But, in equines we generally advise two inseminations at two days interval and that also before the end of oestrus. To reduce the number of inseminations, it is must to know the probable time of ovulation.

Early Detection and Timely Management of Twins

Twin pregnancy is common in many species viz sheep and goat, while it is undesirable in equids. Early detection of multiple ovulation and subsequent management of twins is possible using ultrasonography. In the horse, twin embryos are dizygotic and result from double ovulation. In the mare, if one twin is not resorbed and does not undergo natural reduction to a singleton early in gestation, twin pregnancies usually result in abortion or the birth of non-viable foals, only rarely two healthy, productive offspring produced. Delivery of two foals also places the mare at great risk for trauma and damage of reproductive tract and can even result in death of the mare. These complications are serious enough and so on some farms, a mare with two preovulatory follicles may not have been bred for fear of producing twins.

Early detection of twin pregnancy and crushing of one pregnancy by ultrasonography is extremely helpful to prevent subsequent breeding complications and to improve reproductive efficiency. Manual rupture of one conceptus performed before 30 days of gestation has approximately a 90% success rate in yielding a live foal. The smaller vesicle should be manipulated, separated and crushed at the tip of one uterine horn by monitoring with ultrasonography on or before day 16 because after day 16, the conceptus becomes fixed in position at the base of one uterine horn. If fixation has occurred, one embryo can be ruptured by direct compression of the conceptus with thumb and fingers. After 30 days, management of twins becomes less successful by manual rupture.

Formation and Development of the Corpus Luteum

Corpus luteum (CL) is an endocrine gland, which exists is an intra ovarian structure during two third of mare's estrus cycle and controls important reproductive activities. The presence or absence of the CL can be accurately examined by using ultrasonography than per rectal examination. The corpus luteum is generally detectable with a 5 MHz transducer for approximately 17th day after ovulation (Pierson and Ginther, 1985a). Two distinct luteal morphologies can be distinguished during the dioesturs in mares. Approximately 50% of corpus lutea are uniformly echogenic throughout their period of detectability. The remaining 50% develop a centrally located nonechogenic area representative of a blood clot (corpus hemorrhagicum form). No functional difference appears to exit between the two types of corpora lutea.

Echogenecity or brightness of the CL is greatest during the 24 to 48 hours following its formation. Subsequently, echogenecity decreases until the end of the luteal phase, when it increases again.

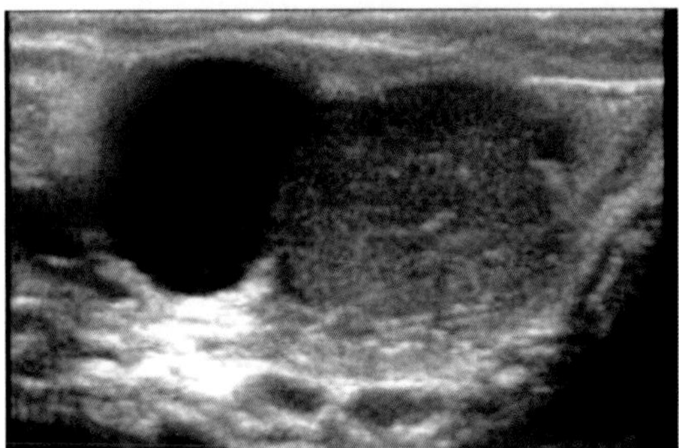

Fig. 76: is showing follicle and corpus luteum on the same ovary

The identification of a CL can be used to determine if a filly has reached puberty or if a mare has entered the ovulatory season, to detect ovulation and to determine if a mare has ovulated more than two follicles.

Early Pregnancy Diagnosis (EPD)

Rectal palpation method had been considered good for pregnancy diagnosis in most of the species including equines. Detection of pregnancy in mares is of considerable economic importance, as it reduces the inter-foaling interval. The primary target of any equine breeder is to get maximum number of foals from any mare in its lifetime. Further, to get one foal a year from a mare, timely diagnosis of pregnancy is essential. Ultrasonography is extremely useful for pregnancy diagnosis earlier than by rectal palpation. Early detection of an embryonic vesicle can be made on day 14-post ovulation (Palmer and Driancourt, 1980). Ultrasonographically, a non-echogenic (black), spherical confirmation of the early conceptus is a peculiarity of the horse and the human, which makes such an early positive diagnosis possible (day 14 post ovulation in the mare and day 20 in the human). The location of the embryonic vesicle in the mare was found to be very constant (in one horn near the body junction) which made it possible to detect the conceptus with in a minute or so after insertion of the rectal probe.

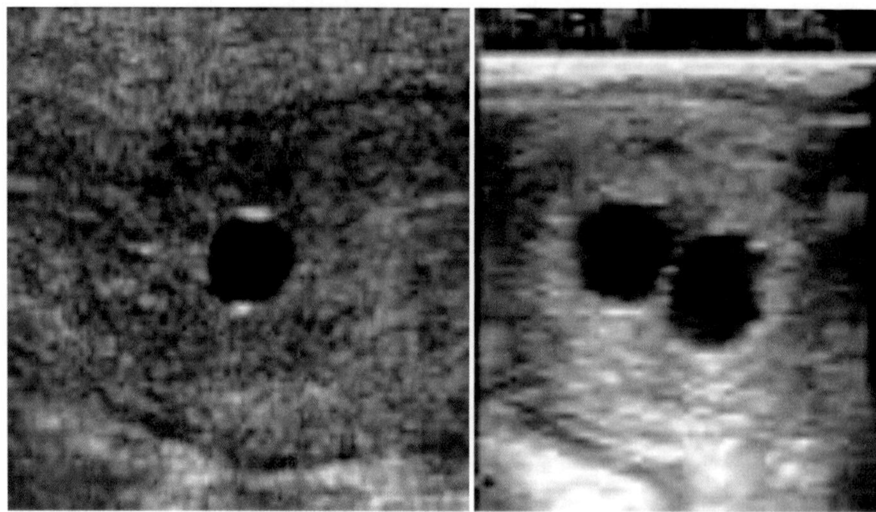

Fig. 77 & 78: showing mare conceptus on day 13 (left) and twins on day 14(right)

In this regard, for early pregnancy diagnosis (before day 15 or 16), it is essential that the entire uterus be scanned, as the vesicle can be located anywhere within the uterine lumen during its mobility phase. Early studies on the efficacy of ultrasonography for early pregnancy diagnosis have demonstrated 97.4% accuracy after day 14. Early pregnancy detection helps in timely recovering of open mares by the stallion.

Foetal Growth and Development

The size of vesicle can be measured by freezing the images on screen during ultrasonography. Until approximately day 16, the vesicle grows at a rate of 3 to 4 mm per day (Ginther, 1986). From day 16 to 28, there is a plateau in the growth of the vesicle and then growth resumes at a slightly slower rate. The embryo proper can be detected by day 20 or 21 and its heart-beat can be detected by day 22 and is an important indicator of embryo's well being (Ginther, 1984). The size and shape of the early vesicle are diagnostically important in estimating the age of a conceptus, particularly when breeding dates are unknown.

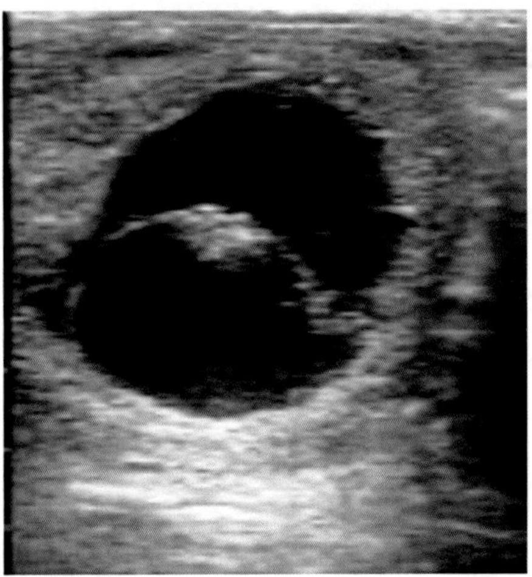

Fig. 79: is Showing mare embryo on day 35

Diagnosis of Early Embryonic Death (FED)

Early embryonic death results in low reproductive efficiency of equids. EED is diagnosed, when an embryonic vesicle is found to be irregular contains fluid in the uterine lumen, echogenic vesicular fluid, no foetal heart beat and slow vesicular grown than normal.

Diagnosis of Post-partum Involution

Ultrasonographic scanning is useful to detect the morphology of the post-partum uterus in mares and time of uterine involution. Involution of the post-partum uterus is achieved within 10-15 days of parturition in equids. A bright, fern-like pattern of hyper-echogenicity outlining the endometrial folds can be observed for an average of approximately two days following parturition. The regularity and echogenicity of this pattern decreased over time. Probably reflecting the stage of involution.

Determination of sex: The technique is also useful for determination of sex of the foetus in pregnant mares on 58-62 days of pregnancy.

Transabdominal Ultrasonography

Transabdominal ultrasonography can be performed in advance pregnancy. In experience of author, from this approach, pregnany and well being of fetus can be detected after 3 months of gestation. Fetal heart and fetal movements can be easily observed. Fetal organs can also be visualized from this position. Fetal eye dimensions can be easily followed for assessing gestation progress.

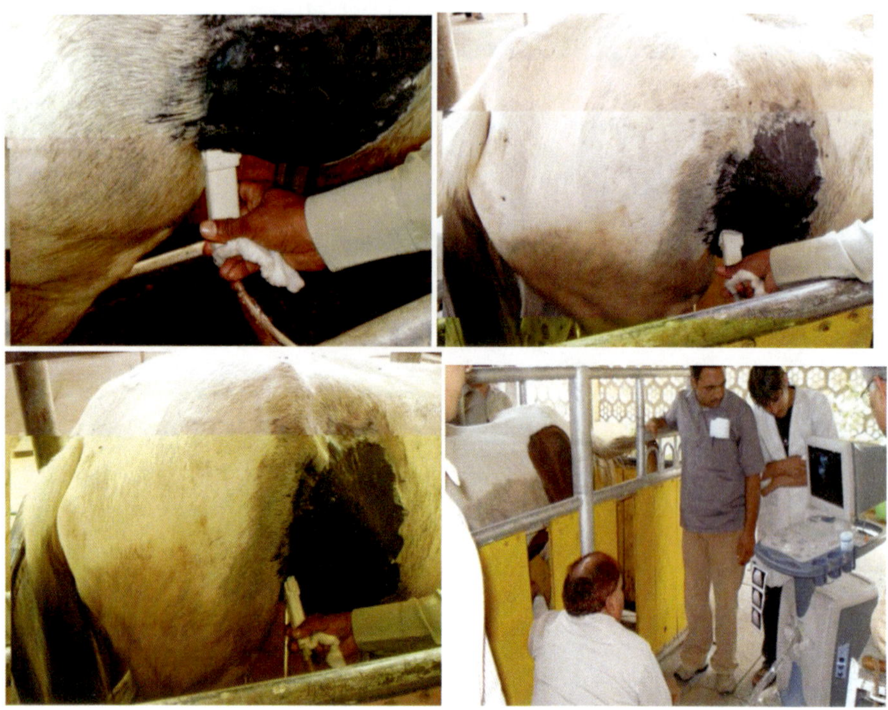

Photograph-21-24: is showing different positions of trans-abdominal ultrasonography in the mare for Pregnancy Diagnosis between 3 – 4 months of gestation

17

Ultrasonic Characterization of Follicle and Corpus Luteum

Corpus luteum, one of the biological clock of the estrous cycle and pregnancy, is known foremost for its production of progesterone that blocks the pituitary release of gonadotrophins and prepare the uterus for pregnancy. The ovarian follicle wall consists of granulosa cell and theca cells vascularised and luteinize after ovulation to from a corpus luteum. Corpus luteum is temporary steroid- producing gland undergoes marked structural and functional changes far a short time span during its development functional life and regression. During estrous cycle, the corpus luteum undergoes an initial period of growth (metestrus), followed by a period of maximal function (diestus) and finally by a period of regression (proestrus) preceding the next ovulation (Robert, 1986). The most prominent cell in the CL is the large luteal steroidogenic cell.

Ultrasound imaging has been used to detect and measure the corpus luteum throughout the estrous cycle and corpus luteum has been characterized during pregnancy and non pregnancy. Changes in corpus luteum diameter as detected by ultrasonography are highly correlated with progesterone production. A high correlation has been found in the ultrasonographic detection of bovine corpora lutea (Lean *et al.*, 1992) and between luteal tissue area measured ultrasonographically and plasma progesterone concentration (Bergfelt *et al.*, 1989).

The B-mode ultrasound scanner transmits and receives high frequency sound waves to create an image of the tissues along a sectional plane corresponding to the plane of the beam path. The section is displayed on a monitor as a 2-dimisional matrix composed of picture elements. A pixel is the smallest single element within an image and can display one of a number of possible shades of gray at one moment in time. The gray-shade of a pixel, which ranges in brightness from pure black to pure white may be represented by a numerical value. The stronger echoes yield brighter pixels and higher pixel values. In B-mode ultrasonography fluids do not reflect ultrasound waves and appear black, whereas dense structure such as bone reflect almost all ultrasound waves and appear bright or white.

Formation of Corpus Luteum

After ovulation, there is enough hemorrhage into the follicular cavity, for a blood clot to develop. The blood filled follicle lumen is without of the ovum is known as corpus hemorrhagicum. This clot of blood serves as physical framework and a nutrient medium for the proliferative granulosa and theca cells. Within 3 to 4 days, the blood clot has been adequately invaded by the new luteal cells so the blood-filled cavity loses its dark coloration. The corpus luteum is one of the most vascular organs of the body. Full size is usually attained by days 8 or 9 of the cycle. The newly formed organ is called corpus luteum verum if the animal becomes pregnant. In the cycling non-pregnant animal, the newly formed organ is termed as corpus luteum spurium. The degenerating avascular non functional corpus is termed a corpus luteum albicans (white).

After ovulation in the cow, granulosa cells hypertrophy and become filled with droplets of a yellow lipid material as the corpus luteum forms

In case of bovine, the maximum size of corpus luteum is attained on day 16 and then it degenerates rapidly. The bovine corpus luteum changes from a light brown to gold on day 7 then to a golden yellow on day 14 and in between days 14 and 20 corpus progressively changes from yellow to orange to brick red. The brick red color remains for several cycles as the old corpus gradually regresses (Mc Donald, 1980).

Ultrasonic Image of Follicle

The ultrasonic image of ovarian follicles characterized by the anechoic, circular area of the follicular lumen because of follicular fluid content usually no reflections. The follicular wall is surrounded by the hyper echoic ovarian stroma can rarely identified. The thin follicular wall is occasionally separated from the ovarian parenchyma by a very narrow hyper echoic line. The shape of follicle is usually round (Pierson and Ginther, 1984). The ultrasonic determination of inner diameter of follicles corresponds quite exact to the real size of the follicular cavity. Ultrasonic images of cavities in cystic corpora lutea resemble those of follicle in contrast to the anechoic follicular fluid; however, the anechoic fluid content of the corpus luteum cavity is surrounded by a moderately echoic wall of luteal tissue, which is few mm thick.

Ultrasonic Characterization of Corpus Luteum

An ovulation only can be detected in a buffalo that was examined by ultrasound in the days preceding the ovulation. On second/third day of ovulation CL, formation can be identified in buffaloes. Newly formed CL is appeared darker on monitor then the older because of tissue of young CL is softer, whereas in case of older the tissue is hard hence, it is more echoic then soft tissue.

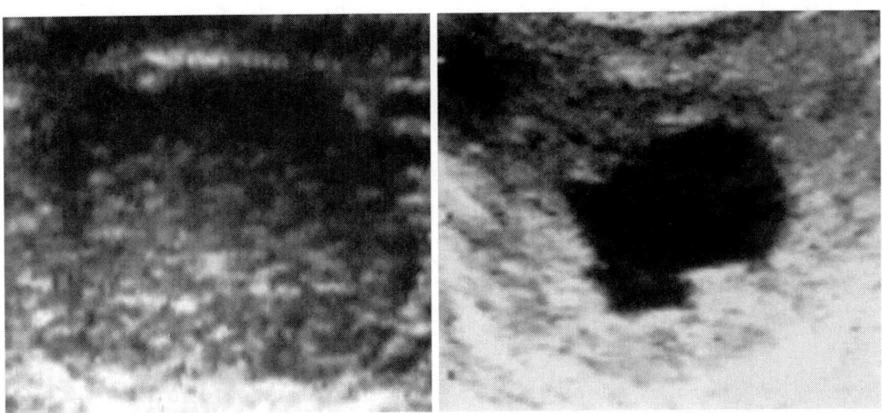

Fig. 80 & 81: are showing bovine Corpus luteum without cavity **(left)** and with cavity **(right)**

The ultrasonic properties of some of the mature corpora lutea were similar to those of ovarian stroma except that the corpora lutea were distinguished by a distinct border. Some of the corpora lutea were characterized by a non-echogenic central cavity, ostensibly filled with blood. The sonograpihic image of luteal tissue appears as a roughly granular gray-structured oval area on the monitor. The hypoechogenicity of the active corpus luteum is in distinct contrast to the brighter gray of the ovarian parenchyma. The latter usually contains several vesicular structures of varying sizes, whereas the luteal tissue contains no fluid except in the case where it developed a cavity during its development. The higher echogenicity of the ovarian stroma, in contrast, reflects its solid consistency and higher tissue density.

The cross section of a corpus luteum contains a narrow, hyper echoic zone in its center. A hyper echoic central zone can always be seen after the closure of a central corpora lutea of the cycle as well as those of pregnancy. The young corpus luteum only becomes sonographically detectable 2 to 4 days post-ovulation. Corpora lutea have a mean width of 14 mm and a mean length of between 18 and 21 mm when they first become detectable in the third day after ovulation (Pierson and Ginther, 1984).

They then grow 1 mm in width and 2mm in length per day and reach their maximum size of about 20x30 mm by day 8 to 10 post ovulation (Kahn, 1994). After regression, CL is still detectable at time of next ovulation. In some cases, CL may be identifiable as corpora albicans for several days after the onset of the next cycle.

The gray scales of the corpora lutea of different ages varies. At the same time, corpora lutea of pregnancy cannot be distinguished on the basis of their echogenicity from those of the cycle. The dimensions of corpora lutea of pregnancy are much the same as those of corpora lutea of the cycle.

Cystic Corpus Luteum

Corpora lutea with cavities so called cystic corpora lutea can also be diagnosed by ultrasonography (Pierson and Ginther, 1984). The cavities inside corpora lutea are usually oval, occasionally round and nearly always centrally positioned inside the gland. Only in exceptional cases are they eccentrically positioned or have they an irregular shape. The largest diameter of the cavities usually varies. The echogenicity of the cavities is similar to that of follicles. The luteinized wall is used to differentiate cystic corpora lutea from ovarian follicles. Whereas follicles are seen to lie embedded in the hyper echoic ovarian parenchyma, the cavity of the cystic corpus luteum is separated from the bright ovarian tissue by its slightly less echoic wall of luteal tissue. As during the estrous cycle, cystic corpora lutea can also be found during the first week of pregnancy.

With the help of ultrasonography, the number and size of corpora lutea on an ovary can be determined quite accurately.

18

Ultrasonography in Camel Reproduction

This paper primarily considers the general principles of ultrasound and instrumentation and its application in camel reproduction.

Ultrasound waves are mechanical pressure waves similar to audible sound waves and must have a medium in which to audible sound waves and must have a medium in which to propagate. The "frequency" (f) is the number of high- or low- pressure regions crossing each area of tissue each second. The frequencies of the sound waves used in medical ultrasound are much higher than the human audible range (20 to 16000 Hz); hence the waves are called ultrasound. Ultrasound instrumentation

The gray scale imaging equipment uses only time delay and echo amplitude information in creating the image. Most equipment contains the same basic building- block circuits; transmitter, receiver, display and scan converter.

Transmitter

The first circuit building block, the transmitter circuit produces either a high-amplitude, short- duration voltage shock pulse or a driving voltage waveform. Then an output control attenuates the amplitude of the shock pulse or the driving voltage wave form before it is applied to the transducer.

Receiver

After the received echoes are converted into weak voltage waveforms by the transducer, they are processed by the next circuit building block, which is called the receiver.

Display

Cathode ray tubes (CRT)

Cathode Ray Tubes (CRT) or broadcast TV tubes are used as the display in most ultrasonic images. The face of the CRT is a two-dimensional surface that can display the echo data (as brightness) in a two-dimensional format.

A-mode (amplitude mode) Display

The amplitudes (A) of the received echoes are demonstrated at the depths of their respective reflectors. This display mode presents the echo amplitude information received from only one transducer line of sight. The horizontal axis is calibrated in distance, which means that measurements of the returning echoes, is converting the measured times into depths in the patient.

B-mode Display (brightness mode)

In this display, the brightness of the B-mode spot on the CRT face is proportional to the amplified echo amplitude. A set of B- mode image lines scanned through a linear plane in the patient (scan plane) is called a B-scan and presents a cross- sectional image of the scan plane called a gray scale image in which the echo amplitudes are encoded in image shades of gray. Gray scale displays present both specula and diffuse echoes for diagnosis. The tissue parenchyma information contained in the low amplitude diffuse echo signals proved to be of great value in diagnosing many different disease status from ultrasonic images. M- mode (motion)

In this mode, the vertical b-mode line is slowly swept in a horizontal direction across the CRT face with a known velocity. The slope of the reflector M-mode (motion) trace gives a quantitative measure of its velocity towards or away from the transducer and the extent of displacement.

Scan Converter

The received echoes are stored in a scan converter (which stores both echo position and amplitude information) and then passed on to the display for viewing and hard copy generation. A scan converter is a memory that accepts data in one scan format and outputs it at a later time in a different scan format. Memory input format is diffused from the display format, which is a broadcast TV format (525 horizontal display lines/ frame at 30 frames/ sec.)

There are two types of scan converters (Goldstein, 1991)

i. ***Analog scan converter:*** presented a gray scale analog image that appeared spatially continuous, much like a photograph.

ii. ***Digital scan converter:*** The echo, spatial and amplitude information are quantified into pixels (picture element) that can be represented by numbers. The received echo amplitudes from each pixel are quantified (analog to digital conversion) and the numbers representing these ranges of echo amplitudes are stored in the computer memory locations that correspond to each pixel.

Real Time Ultrasound

1. ***Mechanical sector scanners:*** The mechanical sectors generally are single element transducers. The transducers are angled back and forth in an oscillatory motion such that the transducer beam defines a pie-shaped sector in the patient cross- section with a limited field of view for anterior structures and a widening field of view for more posterior structures. Some mechanical sector scanners use multiple transducers (usually three) rotating continuously around a common axis.

2. ***Phased array scanners:*** Phased arrays are sector scanners with no moving parts. A set of thin piezoelectric elements are arranged in a line, with their length along the slice thickness direction (perpendicular to the scan plane) and their thinnest dimension along the multi- element array axis.

3. ***Linear arrays:*** A linear array is a linear arrangement of individual elements, but it is much longer and fires in a different manner from the sector array. Linear array form real - time images with a retangular field of view. They are especially useful in obstetrics where their large field of view is useful in studying the fetus and where the large distended abdominal surfaces may be readily coupled acoustically to the entire length of the transducer surface.

3. ***Convex arrays:*** Convex arrays are a variant of the linear array design. They attain large sector type fields of view at depth without side lobe difficulties and a loss of focus at the image edges. They are not useful for imaging at depth.

4. ***Annular arrays:*** Annular arrays are an improvement on focused single crystal transducers. The transducer face is divided into a set of concentric ring element. The rings can be electronically focused in much the same manner as the individual elements in a linear multi element array.

The phased array has the disadvantage of a limited view for anterior structures, but its advantage is that its sector shaped field of view can avoid superficial reflecting structures such as ribs or pockets of bowl gas.

The advantage of the linear array is its large field of view for anterior structures but it has the disadvantage of being affected by superficial reflecting structures and of needing a long contact area with the transducer, which is sometimes impaired by scars or body curvature. Convex arrays have a larger field of view that is useful in many instances.

Real Time

Image generated from reflected ultrasound following sequential activation of the transducer array are displayed on the screen at sufficient speed to give the appearance of a live image.

Piezoelectric Crystals

Crystals of materials such as lead zirconate- titanate, capable of converting applied electrical energy to mechanical deformation and vice versa.

Doppler Ultrasound

When an ultrasound beam meets a moving object the reflected ultrasound is either of increased or decreased frequency, depending on whether the motion is towards or away from the transducer. The Doppler shift frequencies are all within the audible range and are used to drive an audio loudspeaker. Then the Doppler's shift may be analyzed by one of the best computers in the world, the human brain. Its principal use is in peripheral vascular system i.e. moving red blood cells, and cardio- vascular system - constriction of vessel.

Duplex Scanners

It combines a real time imaging mode with a pulsed Doppler range - gated mode using Color Flow Doppler imaging.

Frequency of Transducer

The choice of frequency of transducer is an important issue in obtaining the best scanning results. The lower the frequency of the transducer, the greater the depth at which useful signals can be obtained, but this is counteracted by a loss of resolution. Conversely, a high-frequency transducer will greatly improve resolution but lack penetration. The broad rule, when working intra-rectally, is to use 7.5 MHz for ovarian and early pregnancy studies, 5 MHz for routine pregnancy work over 40 days, and 3.5 MHz for late pregnancy or immediately post-partum. Late pregnancy diagnosis can also be attempted transcutaneously using a 3.5 MHz transducer applied in an inguinal position dorsal to the udder (Boyd, 1995).

Ultrasound Machine

The ultrasound machine used by the authors at N.R.C. on Camel, Bikaner is Scanner - 200 Vet (Pie- Medicals Equipment BV, Phillips wegh. 6227 A J Maastricht, The Netherlands). As per instructions of the manufacturer contrast and gains are adjusted to obtain clear image with good resolution. The machine is placed on a movable trolley at appropriate level for easy visibility of the operator and controls of machines are approachable to operator during scanning.

Probe/ Transducer

For follicular examination endo-vaginal, mechanical and annular array sector probe of dual frequency (5 MHz & 7.5 MHz) while for pregnancy diagnosis linear array dual frequency (5 MHz & 7.5 MHz) rectal probe was used. The frequencies were changed from 5.0 to 7.5 MHz as and when detail of a particular/ structure was desired. But most of the time 5 MHz was used.

The ultrasound machine, video cassette recorder were attached to an electric extension board which was connected to electric stabilizer. Details regarding the identification of she camel, name of the operator were fed to the monitor of ultrasound machine.

Coupling gel, Printer paper & Video cassette

Ultrasound gel was used as coupling gel. Thermal paper of Sony Co. Japan was used to take the print out of desired images. Sony videographic printer, Videocassetts of Panasonic and videocassette recorder of BPL were used for recording of ultrasonographic scannings for the subsequent analysis and re-evaluation.

Technique of Ultrasonography in Camel

Transrectal Scanning

Camel is a large ruminant having normal body weights 550-600 kg and height at withers (2.0 to 2.2 m). Therefore rectal examination of genitalia was attempted in sitting posture. The she camels were restrained in sternal recumbent posture on kutcha but smooth ground and all the four legs were tied. Inj. Xylazine (0.3 mg/kg b.wt.) was administered i.v., which induced sedation in animal within 10 min. and also caused muscular relaxation and effect lasted for 20-30 min. The scanner is placed at a sensible distance from the she-camel, on the side opposite to the operator's rectalling arm. The she - camel's rectum is evacuated of all faeces prior to introduction of the transducer and it is advantageous to carry out a preliminary manual exploration of the topographic examination. The operator's hand was covered with disposable obstetric sleeve and lubricated with soap. The operator took sitting position and introduced the hand gently in the rectum to palpate the ovary and uterine horn. This method gave practically no resistance from the she camel and the genital organs could be palpated thoroughly (Vyas and Sahani, 2000).

The transducer face is lubricated with a suitable coupling medium and is usually covered by a lubricated plastic sleeve and inserted in a cupped, lubricated hand through the anal opening, before progressing cranially along the rectal floor to overlie the reproductive tract. The transducer face is pressed firmly against the rectal mucosa in order to effect ultrasound transmission through the rectal wall into the abdominal viscera.

Urinary bladder was recognized immediately as homogenous non-echoic i.e. complete black and taken as a landmark for genital tract ultrasonography. As the transducer face was moved cranially along the rectal floor, after urinary bladder, the uterus was imaged, as this lies ventral to rectum. When the transducer was moved laterally and slightly downward by changing its angle in the rectum the uterine horns appeared in cross-section to oblique section. Endovaginal scanning for ovarian examination

The endo-vaginal annular array dual frequency (5.0 MHz and 7.5 MHz) probe was covered with a sterile fine lubricated latex condom, with gel (2.5 ml) introduced between the probe and the latex to avoid air gap. The probe was inserted into the vagina and the ovaries were retracted posteriorly with the help of gloved left hand inserted in the rectum. The right hand was used for guiding the probe towards the ovaries. The observations were carried out by same operator recorded on videotape and were subsequently reviewed to monitor the status (growth and/or regression) of individual follicles. Ultrasonographic imaging to monitor early pregnancy in the camel(*Camelus Dromedarius*)

Camel is seasonal breeder and the breeding season of the female camels in India is of short duration (4 months, Dec to Feb) and therefore necessitates the identification of non pregnant females in order to rebreed them with least possible delay. Returns to service are difficult to detect in camels due to induced ovulatory nature and absence of true estrus, and the delay of 60 to 70 days for detection of pregnancy by rectal palpation are the major disadvantages of using these two methods in the camel.

More work has been carried out in South American camelidae like llama and alpaca (Alarcon *et al.*, 1994, Bourke *et al.*, 1992, Gazitua *et al.*, 2001, Mialot and Villemain 1994, Paraguez *et al.*, 1997) than dromedary (one-humped) camel (Skidmore *et al.*, 1992, Tibary and Anouassi, 1997). The use of linear array real time B-mode ultrasound for the detection of early pregnancy in the one-humped camel was tried at National Research Centre on Camel, Bikaner.

Ultrasound examination of internal genitalia is attempted in sitting position as described earlier (Vyas and Sahani, 2000). The female camels are scanned ultrasonographically by the author on 18, 20, 23, 30 (6 camels) and 40 days (3 camels) post mating. The first detection of the embryonic vesicle, embryo proper, heartbeats, allantois and amnion were observed.

The conceptus was recognised at day 18 post mating as a typical 'black-hole' type of fluid accumulation in the lumen of the left uterine horn depending upon the angle of contact of the ultrasound curtain with the uterus. It appeared as discrete and roughly spherical in cross-section and it was irregular and elongated when the ultrasound beam transected the uterine horn tangentially. In one she camel, the examination was done on day 20. The conceptus was found to be elongated considerably and appeared as a discrete and easily recognised accumulation of conceptus fluids. The diameter and outline of this varied appreciably in different parts of the uterine horn. These changes were caused mainly by endometrial folds indenting the conceptus at odd places and by movement of fluid within the conceptus while carrying out the examination.

At day 23 post mating the diameter of the non echogenic conceptus fluid accumulation in the left uterine horn was found to be increased to 2.9 cm (Fig. 4) and the diameter became more uniform along the length of the horn. The embryo itself was first recognizable at this stage as a small echogenic 'blob' inside the conceptus, which is apparently attached closely to the endometrium in the ventral region of the uterus.

At day 30 post-mating, the overall diameter of the conceptus increased more rapidly due to accumulation of fetal fluids. The echoic embryo was imaged lying slightly ventrally within the anechoic luminal centre of the echoic uterine cross- sectional image. As pregnancy advanced the echogenic embryo within the fluid became more prominent as it enlarged steadily and it appeared to detach itself progressively from the uterine wall.

By day 40 post mating the endometrium was found to be raised in irregular manner (folds) throughout its length. A division was observed occasionally between the amniotic fluid surrounding the fetus and the much larger volume of allantoic fluid external to this. The amniotic fluid seems to contain echogenic debris, whereas the allantoic fluid was much cleaner and non-echoic.

The fetus was easily recognizable as having identifiable features by day 40 post- mating. The head, trunk and limb regions were clearly visible. The optic area was imaged as non-echogenic or anechoic spot in the head region. The ultrasonographic appearance of the fetus and fetal parts depended greatly on the orientation of the fetus and that of the transducer. This clearly indicates that at day 40 post-mating, accuracy of fetal parts identification and measurement is difficult and is influenced by movement of animal, uterine horn, conceptus fluid, positioning of the transducer and the plane of the ultrasound wave cutting across the uterus.

The small hypoechoic dot within the embryo's image observed on day 30 post mating was the region of the embryonic heart. At day 30 post mating the embryonic heartbeat was discernible in real-time as a rapid rhythmical fluttering movement in the cardiac region located in centre of the echogenic embryo. In the present study heartbeat was counted for 15 seconds and then multiplied by 4 to take the values of beats per minute. The mean value of heartbeat on day 30 and 40 was 202.7 ± 3.5 and 174.7 ± 3.5 beats/min respectively.

The corpus luteum is the major source of progesterone during pregnancy in one humped camels and its presence is required throughout the gestation period. Therefore in the present study, ultrasonographic visualization of corpus luteum was also taken, which added to the accuracy of the early pregnancy diagnosis. The corpus luteum of pregnancy was observed to be larger in diameter and it was observed to have a central hypo- or non-echoic fluid filled area on day 20, 30 and 40 post-mating.

Ultrasonography is a non-invasive and powerful tool for evaluating early pregnancy diagnosis, fetal heart-rate and placental development in the human (Birnholz, 1984), as well as in domestic animals like cattle (Curran *et al.*, 1986a & Curran *et al.*, 1986b), mare (Ginther 1983), sheep (Fowler and Wilkins, 1984), pigs (Martinat-Botte *et al.*, 1998), llama and alpaca (Mialot and Villemain 1994, Paraguez *et al.*, 1997, Gazitua *et al.*, 2001).

Ultrasound technology can also prove to be a reliable tool for detecting a conceptus as early as day 20 post mating in the camel, since return to service is difficult to detect in the camel, due to lack of definite estrous cycle and less obvious external signs of estrus. Some information is available regarding the early detection of camel embryo and fetal cardiac activity in the literature (Tibary and Anouassi, 1996).

It was demonstrated in the present investigation that ultrasonographic examination allows accurate diagnosis of pregnancy in the dromedary camel at a much earlier stage of gestation than is possible by palpation alone.

The first discrete accumulation of fetal fluids in the uterine lumen was visible by day 18 after mating. Observations in the present study by real time ultrasound of the pregnant she camel allowed for the visualization of the embryonic vesicle on day 18. The vesicle was first spherical and became elongated and irregular in shape after day 23. But, when the transducer was placed on ventro-lateral aspect of the horn the embryonic vesicle was visible as spherical. Definite diagnosis of pregnancy could be made on the basis of accumulated fetal fluids and the echogenic embryo with its pulsatile heart, by day 30 after mating. Similar results but with early diagnosis of fetus on day

20 post ovulation (Tibary and Anouassi,1997, Tinson and Mckinnon, 1992) and day 23 post ovulation (Skidmore *et al.*, 1996) were reported earlier. The time elapsed between mating and ovulation in camel is 24 to 36 h (Marie and Anouassi, 1986). Therefore, difference in day of first detection of embryo proper in the present study and the earlier reports is not long. However this is at least 30 days before pregnancy in she camel can be diagnosed confidently by rectal palpation (Banerjee *et al.*, 1981).

In the present investigation the amniotic fluid owing to presence of slightly echogenic debris could be discerned easily from the allantoic fluid, which was much clear, larger in volume and completely non-echoic. But the echogenic amniotic band surrounding the fetus, which is characteristic of bovine (Boyd 1995) and buffalo fetus (Pawshe *et al.*, 1994) throughout the gestation, was not observed in the present study. However the present findings are similar to earlier reports (Skidmore *et al.*, 1992, Tibary and Anouassi, 1996 and Tinson and Mckinnon, 1992) that did not mention about echogenic amniotic ring or band around camel conceptus.

Use of realtime ultrasonography for evaluating follicular activity during the non-breeding season (summer months)

The one humped camel (*Camelus dromedarius*) is a seasonal breeder with a relatively short breeding period, when increased ovarian activity is exhibited. The breeding season varies in different parts of the world depending on the geoclimatic conditions of the region viz. December - March in Pakistan, April to May in Somalia, March to August in Sudan, February to March and August to September in Mali, May to June in Morocco and December to August in Egypt (Wilson, 1989). In India, the breeding season of the camel extends from December through March (Khanna *et al.*, 1990). Because of the 13 months gestation period, the post partum period and the periods of anoestrus associated with lactation and season, mating and calving periods coincide. Thus camels, which calved will not normally be mated until the following breeding season, giving an inter-calving interval of two years or longer (Minoia *et al.*, 1992). The exogenous hormonal therapy to breed the camels during non-breeding season has been used with moderate success (Elias *et al.*, 1985, Agarwal *et al.*, 1997).

The present study was conducted (i) to study the ovarian activity in she camels using ultrasound examination (ii) to explore the possibility of breeding the she camels during the June - August which is not considered to be breeding season for camel and (iii) to monitor ovulation and pregnancy in mated camels by ultrasound.

The experiment was conducted during June to August (peak summer and initial monsoon) when maximum ranged from 40 to 47 ^{0}C and the duration of sunlight was 14 h.

Sixteen adult (age, 8-10 yr), pleuriparous, non-pregnant, female one humped camels (Camelus dromedarius) belonging to herd of National Research Centre on Camel, Bikaner, located in Thar desert, India, situated 28.3 0 North latitude and 73.5 0 East longitude were taken for the experiment. The camels were mated with virile studs during previous breeding season but remained non-pregnant. The animals were maintained under semi-intensive management system. In this system, camels were sent out for grazing on rangeland from 9.00 AM to 4.00 PM and were offered dry fodder @ 10 kg per animal in the evening.

The male camels were housed next to the experimental females during the experiment. To check the libido the males were paraded among females twice daily for 30 min in the morning and evening (6.00 and 19.00 h).

Ovarian examination

The endovaginal annular assay probe (5 MHz) of ultrasound scanner-200 (Philips Medical System, Delhi, India) was used for ovarian examination (Vyas and Sahani, 2000). The animals were examined at weekly interval during these months. All follicles ^{3}0.5 cm in diameter present on both the ovaries were counted and measured using the internal electronic calipers. The females having follicles of ^{3}1.0 cm diameter were allowed to be mated. The male camels showing symptoms of rut were used for breeding. The females having follicles of > 1.0 cm diameters were mated. The females were mated at early morning (6 AM) and after 24 h (Vyas *et al.*, 2004).

The ultrasound examination of the ovaries revealed presence of ovulating size follicle (> 1.0 cm) in 8 out of 16 camels. No follicular activity was observed in the rest of the 8 camels during the period of experiment. Three females possessing follicle of ovulating size were observed to mount over other females. But it was not observed in rest of the five females. Three out of 10 adult males expressed symptoms of rut and were used for breeding. Four out of eight camels having mated with virile males became pregnant which was confirmed by ultrasound examination.

The results of the present investigation revealed that follicular activity do occur in about 50 % of the non-lactating she camels during the months of June to August, which are generally considered to be part of non-breeding season. However camel owners do not breed and prefer calving during this period due to high temperature and humidity stress.

Sghiri and Driancourt (1999) reported a peak of calving (over 80-90%) among dromedary camels in Morocco from December to April and hardly any calving between June and October. In India, the breeding season of the camel extend from November to March as calvings occurred predominantly between December and march (Khanna *et al.*, 1990). A number of previous reports have also shown a clear breeding season in a number of countries where clear changes in photoperiod and/or food supply occur throughout the year (Egypt: El-Wishy and Ghoneim, 1986;United Arab Emirates: Aboul Ela, 1994; Saudi Arabia; Abdel Rahim *et al.*, 1994).

The present findings add to the above findings in showing that follicular activity can persists for 2-3 months beyond the normal limit of the breeding season in some of the those females which failed to conceive during the breeding season. It was also revealed in the present experiment that clear behavioural symptoms of follicular phase are lacking in majority of female camels, which is in agreement with findings of previous workers (Skidmore *et al.*, 1996a). In conclusion, our study showed that ovulating size follicles are observed in some female camels during June to August (non breeding season in India) and successful pregnancies may be achieved after mating of individual animals during this period.

Real time ultrasonography of ovaries during the early postpartum period

The involution of uterus following parturition is reported to be complete by 21±0.5 days (Musa and Makawi, 1989) or 39 to 42 days (Ahmed, 1990) postpartum. In the present study ovarian status was examined by real time ultrasonography with an aim to explore the possibility of breeding camels during the early post partum period for improvement of reproductive efficiency.

Adult pleuriparous, lactating she camels (n=17) belonging to the herd of National Research Centre on Camel, Bikaner, located in the Thar Desert, India and parturited in the first half of breeding season were included in the present study. Each she camel had a suckling calf. The endovaginal annular array probe (5 MHz) of ultrasound scanner-200 (Philips Medical System, Delhi, India) was used for ovarian examination (Vyas and Sahani 2000). The camels were examined at 30, 45, 60, 75, and 90 days post-partum. All follicles were searched on both the ovaries and measured using the internal electronic calipers. When a follicle of ^{3} 0.5 cm diameter was observed on an ovary, the animal was selected for frequent examination at a shorter duration of twice weekly to monitor the growth of the follicle. The presence of a follicle ^{3}1.0 cm diameter in either of ovary was the criteria used to decide the mating of a camel with a virile stud. The mating was repeated at a 12 h interval and hCG (5000 i.u., Profasi, Serono, India) administered intramuscularly to facilitate a higher rate of ovulation. The ovulation was assessed by ultrasound examination at 72 and 96 h post-mating. The involution of uterus as assessed by recto-genital palpation was completed in 25 to 30 days in every camel. Ovarian scanning upto 90 days postpartum revealed not a single follicle in either of the ovaries in 5 (29.4 %) of the camels. Atleast one follicle of ^{3} 0.5cm diameter was found, between 30-60 days postpartum in the remaining 12 (70.6 %) camels. In three camels a follicle ^{3}0.5 cm diameter was found by day 45 postpartum. However no further increase in follicular size was observed on repeated ovarian examination in three cases. In other 9 (52.9%) camels, further increases in follicular size were observed, and atleast one follicle of ^{3}1.0 cm diameter was found between 34 to 70 days (mean=56.6±3.4 days) postpartum. The administration of hCG 5000 i.u. intramuscularly at the time of mating induced ovulation in 100 % of cases between 72 and 96 h of the mating. Pregnancy diagnosis at 60 days post-mating showed that 4 of these 9 mated camels conceived for an overall conception rate of 44.44%.

In this study, it was impossible to relate estrous behaviour shown by the female camels to follicular activity in their ovaries. The presence of a dominant growing follicle was not always associated with clinical evidence of estrus or sexual receptive behaviour in camels. More often than not, camels having a follicle of ^{3}1.0 cm diameter were mated when they appeared to resent mating. Conversely, females with no follicular activity in their ovaries were showing clear signs of estrus, like frequent urination, stradling of the hind leg, mounting other females, and submissive behaviour towards an approaching male. However they were ignored by the male.

Follicular growth was observed more commonly in young camels. Follicles of ^{3}0.5 cm diameter were found in 8 out of 9 (88.9 %) camels and follicle of ^{3}1.0 cm was observed in 6 of 9 (66.7 %) camels in this age group. The 4 camels which conceived were from this age group. In older age group (11-20 yr), a follicle of ^{3}0.5 cm diameter was found in 50 % of animals, and a follicle of ^{3}1.0 cm diameter could be observed only in 3 of 8 (37.5 %) camels.

Postpartum resumption of ovarian function and follicular activity is highly variable in camels (Tibary and Anouassi, 1997). The first postpartum estrus was reported to be delayed for one year (Yasin and Wahid, 1957, Matharu, 1966 and Williamson and Payne, 1978 and Wilson, 1989) and irrespective of date of birth, until the next breeding season (Musa and Makawi, 1989). In another study, a variable number of mature follicles on the left ovary of non-suckled animals was observed at 39-42 days postpartum; 3 camels were mated but none conceived (Elias, 1990). No growth of mature follicles (Ahmed,

1990) and an interval of 4.5 to 10 months for first postpartum estrus (Evans and Powys, 1979) was reported in lactating dromedaries. In the present study, 52.9 % of the lactating and suckled animals were observed to have follicle of ^{3} 1.0 cm diameter. It would appear that lactation may not have a negative effect on folliculogenesis in the camel. However, the hormonal mechanism that drive the development of follicles, and the reason why they occur in only a proportion of postpartum camels, have yet to be elucidated. A shorter duration of 45 and 25 days to first postpartum estrus is reported in a small study on stall-fed lactating camels (Yagil and Etzion, 1984) indicating a possible influence of improved feeding and management on the postpartum estrus. In the Bactrian camel (Camelus bactrianus) Chen and Yuen (1979) described only follicular (1.0-1.4 cm) growth in 33 % animals and no conceptions. The percentage of animals having follicular growth and conception rate in the present study was higher than the results in bactrian camels reported by previous workers.

In the present study, it was observed that the resumption of follicular activity after parturition was better (88.9%) in younger camels (8-11 yr), than older camels (11-20 yr, 50%). The conception was also better in younger camels. The involution of uterus as assessed by size and shape, was complete in both age groups, but insufficient hormonal and endometrial milieu could be the probable reason for lower rate of follicular activity and conception in older camels.

An estrous period of 3 to 6 days (Joshi *et al.*, 1978, Elias *et al.*, 1984) to 21 days (Yasin and Wahid 1957) with definite signs of estrus has been reported. However signs of estrus in this study were observed to be highly variable and had a poor relationship with postpartum follicular activity.

These findings confirm the earlier reports in one humped camel (Skidmore *et al.*, 1996) and llama (England *et al.*, 1971) that signs of estrus bear little relationship to follicle maturity. In summary, postpartum follicular (^{3}1.0 cm diameter) activity is observed in some female dromedaries. If mated, pregnancy can be achieved at 34-70 days of parturition in some of these cases. This kind of breeding management could reduce the calving interval significantly in the camel.

Ultrasound Scanning to Examine Follicular Cycle

The folliculogenesis was studied using annular array dual frequency endovaginal probe of ultrasound scanner-200 during breeding (February) and non-breeding (September) season (Vyas and Sahani, 2000). All follicles ≥0.5cm in diameter and corpora lutea present in both the ovaries were counted measured using internal electronic calipers. The observations were recorded using video cassette recorder interfaced with scanner.

An adult she came having normal genitalia and calved atleast once was selected for daily scanning of both ovaries for a period of 30 days. The she camel was not mated during the experiment.

(a) Non-breeding Season

Left ovary: On the first day of the examination three follicles were present on left ovary. One of the follicle increased in size from 1.66 cm on subsequent on 9th day and it continued to grow up to day 15th acquiring 4.3 cm size. The observations on 15th day onwards exhibited regression of the follicle, which was traceable upto 24th day only. The other two follicles (size 1.15 & 0.4 cm) did not reveal any change in the size till 17th day of examination but a slight increase was noticed in one of the three on day 20.

Right ovary: One follicle (approximately 0.6 cm) was present on right ovary on first day of examination. Subsequent examination revealed that the follicle did not grow and disappeared. This study revealed folliculogenesis during non- breeding season in she camels with follicular cycle in vogue.

(b) Breeding Season

The experimental she camels used for ultrasound scanning during non-breeding season were not mated between September and February. The investigations on folliculogenesis revealed:

Right ovary: In all three follicles measuring more than 0.5 cm could be located on right ovary during ultrasound examination. On day 1st of the examination only one follicle was located on right ovary. The follicle measured 2.59 cm. A phase of gradual increase of this follicle started again on day 17th and it attained a size of 2.90 cm on day 26th. It declined thereafter reaching 2.22 cm on day 30th. The second follicle (0.8 cm) was found on day 6th of the examination. It did not change markedly upto day 24th and grew thereafter to 1.99 cm on day 30th. A third follicle (0.8 cm) was also observed between day 20th to day 27th.

Left ovary: On left ovary also 3 follicles (0.5 cm) could be found during this period. A mature follicle measuring 1.68 cm was present on first day of examination. It grew upto 4.0 cm on day 13th and then the follicle regressed upto day 17th. The follicle become atretic therafter and could be traced only upto day 27th. A small follicle was found between day 2nd to day 6th, but did not exceed beyond 1.0 cm. The third follicle was observed from day 11th to day 25th. This also did not grow beyond 1.30 cm during the period of observation.

The results indicated that apart from breeding season folliculogenesis also occurred during non-breeding season. Follicular growth and regression is a gradual and sequential process in absence of ovulation, similar findings have also been reported by Skidmore *et al* (1996). It was also observed that only one follicle grew at a time to mature, the growth of other follicles followed the process of regression of the matured follicle which is consistent with follicular wave theory proposed for cows by Ginther *et al* (1989) and dromedary camel by Skidmore *et al* (1997).

Ultrasound monitoring of females for follicular growth during different part of breeding season.

In November 28 U/S exam on 18 females and was performed. In first week of December 7 females were examined. In November 9 (50 %) females had ovulating size follicle. And one female had smaller follicle (0.7 cm). In first week of December 6 out of seven females possessed ovulating follicle and were bred. Four females became pregnant in first service. During breeding season, in the month of January, 29 U/S exams over 20 females revealed that ovulating size follicle was present in 18 females. In February all females examined possessed ovulating follicle. In March ovulating size follicle was observed in all five examinations over five females.

The results revealed that during early phase of breeding season the follicular activity was not present in all camels. In mid breeding season almost all breedable females had ovulating size follicle in one or both ovaries. Once the follicular activity starts it persists till the end of breeding season and even beyond. Therefore ultrasound examination of female camels is advised especially during the early breeding season to identify ovulating follicle. Breeding of females during early breeding season help the breeders to plan rebreeding of these females during post-partum period when they deliver the calf in the next breeding season.

19

Ultrasonography of Mammary Glands in Buffaloes

Mammary glands are important part of body which synthesize and store the milk in between milkings or suckling by the calf. Therefore it attains prime importance in any dairy animal. However, it has received little attention in buffaloes till now. Buffalo presents several unique characteristics in terms of milk production, length and shape of teats and arrangement of duct system in the mammary glands. Development of udder may be related to milk production and the knowledge of changes in its structure in relation to different physiological stages may also have practical implication.

Transducer Selection and Preparation of Scanning

Preparation of the site to be imaged will enhance the quality of the image. Therefore, any loose hair and dirt must be clean off. Ultrasound gel is applied to the area of interest. The gel acts as a coupling agent to prevent air from trapping between the transducer and skin. A 7.5 MHz transducer is ideal for imaging the teats. A 5.0 MHz transducer provides an adequate image. The mammary gland should be scanned with a 5.0-7.5 MHz transducer to give adequate penetration. All structures should be scanned in cross section and longitudinal planes.

Anatomy and Physiology of the Buffalo Udder and Teat

The buffalo has an udder similar to the cattle in the gross anatomy. The buffalo has four teats and some time extra teats also found. The teats are varying in shape and size. Generally, they are larger than cattle teats. Cylindrical forms of the teats are common in the Murrah breed. The hind teats are longer and have larger diameter than the fore teats. The hind quarters of the udder are slightly larger than the front ones and contain more milk. The approximate ratio is 60:40 (hind: front), as for cattle. It takes a longer time to milk the hind quarters.

The Mammary gland is made up of alveoli and alveolar ducts. The alveoli are grouped into lobes and lobules. The small collecting ducts open into large lactiferous ducts which form a pocket/pouch like structure before entering into the gland cistern or lactiferous sinus. The gland cistern is separated from the teat cistern by a fibrous annular folds/ring which is 8-12 in numbers. The wall of the teat cistern is composed of 5 layers: the outer stratified squamous epithelium, muscular layer with longitudinal and circular layers, connective tissue which contains the major blood supply, the sub-mucosa and the inner mucosa. The streak canal is closed by a sphincter of smooth muscle and elastic tissue. The junction of the mucosa of the teat cistern and the stratified squamous epithelium of the streak canal is the rosette of Furstenberg.

The anatomy of buffalo teats is slightly different from cattle teats. The epithelium and sphincter muscle of the streak canal is thicker in buffaloes than in cattle as a result greater force is required to open the streak canal. Therefore, buffaloes are said to be "hard milkers".

Milk is synthesized in the alveoli and stored in upper glandular part of the udder (alveoli and small ducts) between the two milkings. Hence, buffaloes have no cisternal milk fraction. The milk is expelled to the cistern only during actual milk ejection. Because of the absence of cisternal milk between milking, in the teat cisterns, the teats are collapsed and soft before let down. This is contradictory to the cow, where the teats can be very hard and firm due to the presence of milk in the teat cistern.

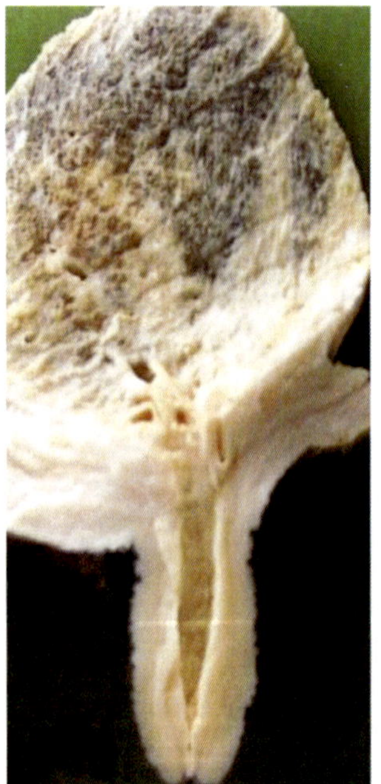

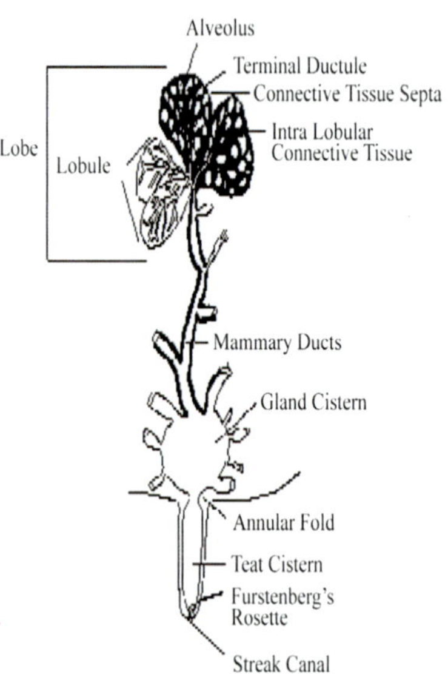

Photograph-25: is showing Longitudinal section of mammary gland

Ultrasonographic Appearance of the Normal Buffalo Mammary Gland
Ultrasonography can be performed in standing animal using intra-operative real- time 5.0 and 7.5 MHz convex probe connected to real time ultrasound equipment (SSA-220A, Just Vision, Toshiba). Mammary quarters are examined directly and teats are examined under water bath. Ultrasound gel is used as a contact medium. The body of the udder is examined using a caudal approach at three positions in both vertical and horizontal cross sections; at the base, in the middle, and proximal to the gland cistern. The distal part of the gland cistern, the teat cistern and the teat canal are examined in vertical/ horizontal cross section. To optimize images of the glandular parenchyma, the probe is kept directly on the skin after gel application. Examination of teats is conducted through semi transparent latex cone filled with water. Lubricating gel is applied on the probe to improve contact with the water cone. The probe is manipulated until a clear image of the teat appeared on the screen. When the picture is obtained, the image is frozen on screen. Printout of static images of the mammary gland and teats are taken on thermal paper.

The glandular parenchyma appears as a mixed trabecular pattern of anechoic (black) areas compartmentalized by hyperechoic (white) partitions. The anechoic areas represent milk within the gland. Small ducts in the matrix of the gland lead into larger lactiferous ducts which formed a pocket like structure before entering into the gland cistern. The pattern formed by the lactiferous ducts and glandular parenchyma may vary among individuals. Non-lactating mammary glands appear as dense hyperechoic tissue with few visible blood vessels and lactiferous ducts. Blood vessels are seen within the mammary gland, but they are difficult to differentiate from lactiferous ducts. The gland cisterns appears as a large anechoic area. Buffaloes have small gland cisterns area. The teat parameters measured may includ teat canal length, teat diameter, cistern diameter, and teat wall thickness. Teat diameter is measured across the distal end of the teat canal. The diameter of hind teat cistern (25.83±1.46 mm) found to be significantly ($p>0.001$) higher than fore teat cistern (16.29±1.12 mm). The annular folds (11.26±2.53 mm) could be detected as a hyper echoic band of tissue separating the gland cistern from the teat cistern. The teat cistern has an anechoic lumen surrounded by 5 layers visible on ultrasonographic examination. The outer layer represents the skin air interface. The next inner layer is intermediate in echogenicity and represents the muscular layer. The 2 layers of longitudinal and circular muscle are difficult to differentiate on ultrasonography. The next inner layer is thin and hypoechoic to anechoic and represented the blood vessels known as the plexus venosus papillaris. The innermost layer is hyperechoic and represents the submucosa and the mucosa. Teat wall thickness at middle (7.45±0.45 mm) is significantly lower than apex

(13.21±1.09 mm) of teat. The teat canal appears as a thin, hyper echogenic line with two neighboring, parallel and thick, hypoechogenic bands. The teat canal can be measured directly on the sonogram.

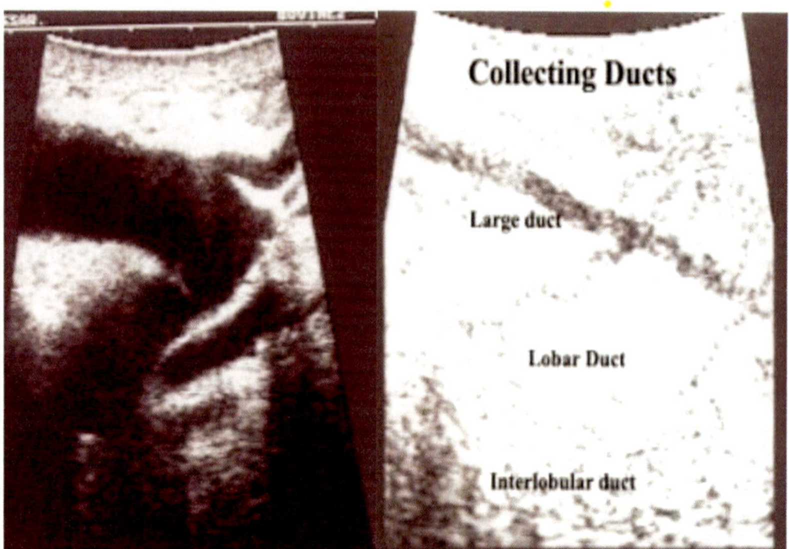

Fig. 82-83: are showing USG image of udder depicting milk vein **(left)** and udder showing drainage system **(right)**.

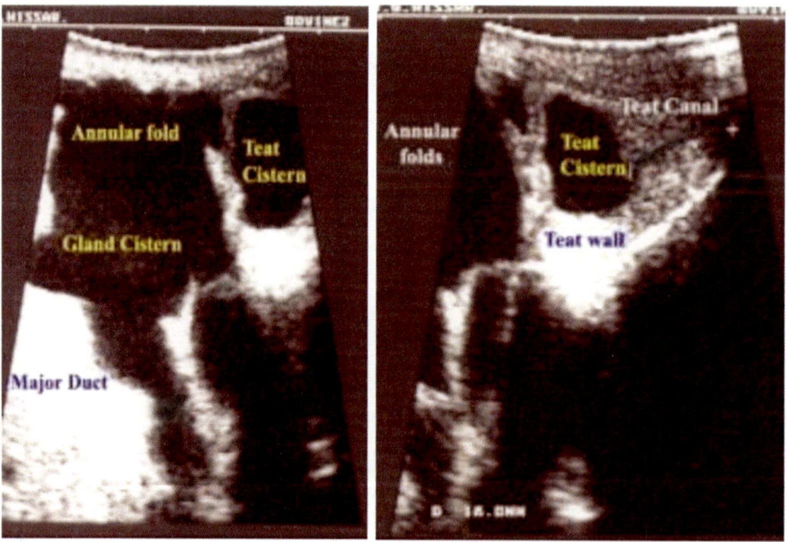

Fig. 84-85: are showing images of udder showing gland cisterna and annular folds **(left)** and teat cistern and annular folds **(right)**

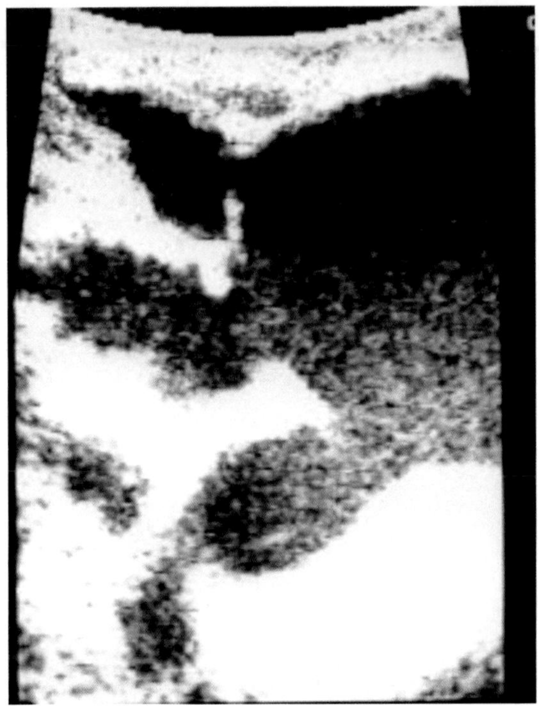

Fig. 86: is showing ultrasonogram of udder showing pocket formation before gland cistern

The gland cistern appears as a large anechoic area. The lining of the gland cistern appears hyperechoic. The annular ring can be detected as a hyperechoic band of tissue separating the gland cistern from the teat cistern. The teat cistern has an anechoic lumen surrounded by 4 layers visible on ultrasonographic examination. The most outer layer represents the skin air interface. The next inner layer is intermediate in echogenicity and represents the muscular layer. The 2 layers of longitudinal and circular muscle are difficult to differentiate on ultrasonography. The next most inner layer is thin and hypoechoic to anechoic and represents the blood vessels known as the plexus venosus papillaris and the circulus venosus papillae. The innermost layer is hyperechoic and represents the submucosa and the mucosa. The muscular layer of the streak canal appears hyperechoic.

Conclusion

Currently doing reproductive ultrasound, it is very simple to make mammary gland structure and anomaly diagnosis. Ultrasound diagnosis can provide a more definitive diagnosis on those cases where the veterinarian is just not 100% sure. Use of Ultrasonography is very useful to measure udder and teat canal characteristics in buffaloes. Ultrasonography appeared to be a useful additional diagnostic tool in evaluating the health of the udder of small ruminants.

20

Ultrasonography for Diagnosis of Mastitis and Teats Abnormalities

Ultrasound is mainly used for reproductive purposes in bovine, now days it is also used for other applications. Ultrasound is an excellent tool for diagnostic imaging in large animals because it is portable, easy to use in the field, produces results at the time of the examination. It is totally non-invasive. Palpation of a gland can only detect certain levels of abnormality. Ultrasound can give an actual perception of mammary health, a picture of the inside anatomy and pathology. Tumors, scarring, stones, trauma, hematoma formation and level of damage due to mastitis can all be evaluated by this technique. Other diagnostic tools are still primary diagnostic tools, ultrasonogram gives yet another dimension to the hard to diagnose case. It will also allow a better estimate of treatment prognosis to aid in deciding whether to cull or treat a dairy animal.

Congenital anomalies at initial stages of the mammary gland development during gestation may result into presence of supernumerary teats or less number of external teats, in place of usual presence of four teats in buffaloes. Hence, cases with two / three / five or six teats are often encountered in the field.

Ultrasound examination of the mammary gland can confirm diagnoses of tissue destruction, scarring, non-palpable stones (calcification). Ultrasound transducers intended for rectal examination of the reproductive system are ideal for examination of the mammary gland and teats. Ultrasonography of mammary glands is helpful in early diagnosis of mastitis or any deviation in udder and teats structure.

Buffalo have four mammary glands which drain with four teats. However, some buffaloes may also have four mammary glands opening into two and three external teats. In the buffalo having three teats, ultrasonography revealed that complete udder was having four mammary glands anatomically while morphologically it looks like three mammary glands. Two mammary glands with two gland cisterns and two teat cisterns opened through the single teat canal containing in single teat visible externally.

Mastitis

Mastitis is an inflammation of one or more quarters of the udder usually caused by bacterial infection. Several types of bacteria cause the different mastitis infections. Mastitis may also be caused by trauma such as injuries on the udder or the teat sores are a perfect way of entry for bacteria. It cannot be eradicated but can be reduced to low levels by good management of dairy animal if it is identified at an earlier stage. The susceptibility of animals varies considerably and new infections are most common in older animals when the management is poor. The disease causes great economic losses both for farmers and for dairies.

Mastitis reduces milk yield, increases the cost of production and makes milk less valuable.

Mastitis can be clinical and sub-clinical. There is an important difference between these two. Sub-clinical mastitis may not even show. The observant milker may see slight changes in the milk like the appearance of flakes, when fore milking in the strip cup. Sub-clinical mastitis shows in the CMT.

Acute mastitis is detected very easily. The udder is sore, swollen, hot and red. The buffalo is in pain when the inflamed quarter is touched. An inflamed quarter should be milked by hand very carefully, not forcing the milk out. This may be quite difficult, since the milk is thick (more like jelly) and may contain blood. The milk should not come into contact with any containers used for milking, nor animals or the floor. It should be wasted. The milker should carefully wash his hands before touching any other quarter, animal or equipment.

Buffaloes have a powerful defense against mastitis due to the anatomy of the teat. Starting with the teat skin, the buffalo teat skin is less sensitive than cow teats to chapping and sores. Inside the streak canal, the epithelium is thicker and more compact than in cattle. This gives extra resistance against penetration of bacteria through the epithelium. Furthermore, the keratin layer of the streak canal is thicker. The importance of the keratin layer is that it contains bactericidal and bacteristatic lipids and cationic proteins. The cationic proteins are inhibitory to the growth of Streptococcus agalactica and Staphylococcus aureus.

Ultrasonography of Mastitis Affected Udder

Acute mastitis is denoted by marked udder enlargement with engorgement of the lactiferous ducts with fluid. The cell content of the milk is high. Ultrasonography of the ducts appeared as hypoechoic due to the increased particulate matter. The demarcation between fluid within the lactiferous ducts and the glandular parenchyma may be diminished due to edema within the gland. A discrete abscess may be visualized anywhere within the glandular parenchyma as hypoechoic surrounded by fibrous wall. Chronic mastitis showed more fibrous tissues within the glandular parenchyma and was decreased in the lactiferous ducts, and whole gland appeared as hyperechoic. Calcification appeared echoic in gland cistern. Ultrasonography showed that teat cistern was replaced with fibrosis in the mastitis affected blind teats. No anechoic milk was visualized within the teat cistern. In the mastitis cases the teat block started from base of teat or towards teat canal. Chronic diffuse mastitis results in an increase in fibrous tissue within the glandular parenchyma and a decrease in the lactiferous ducts, giving the gland an overall hyperechoic appearance and formed a web within the teat cistern causing teat obstruction.

Ultrasonography showed that teat cistern was replaced with fibrosis in the mastitis affected blind teats and it becomes very short (6.77 ±1.42 mm) than the normal rear teat (22.83±2.30 mm). Teat wall thickness also increased (10.80 ± 0.45 mm) in the mastitis affected teats than the normal teat (7.45 ± 0.45 mm) of the same animal. No anechoic milk was visualized within the teat cistern. In the mastitis cases the teat block started from base of teat or towards teat canal. Chronic mastitis formed a web within the teat cistern causing teat obstruction

Obstruction of the Teat

Obstruction of the teat may occur as a congenital anomaly or secondary to a traumatic occurrence. The obstruction may occur anywhere from the teat orifice to the gland cistern. Ultrasound can be used to determine the location of the obstruction and its extent. If the teat is completely obstructed the teat will be filled with milk proximal to the obstruction. Distal to the obstruction the teat will appear collapsed and docs not contain milk. To further delineate the thickness, structure and extent of the obstruction, saline can be infused retrograde into the teat canal. This technique will outline the obstruction and provide information to obtain a treatment plan and prognosis. In the case of a membranous shelf, once filled with saline the teat cistern may have normal anatomy and the proximal obstruction can be defined. Obstruction of the teat cistern may also be due to complete fibrosis. No anechoic milk will be visualized within the teat cistern instead it will be replaced with hypo to hyperechoic dense tissue. A teat cannula cannot be passed in this case. Trauma or chronic mastitis may form a web within the teat cistern causing teat obstruction. The teat cistern will be filled with anechoic milk but thin hypoechoic strands can be visualized emanating from the walls of the teat cistern. If the streak canal is obstructed, the teat cistern will be filled with anechoic milk to the streak canal with normal anatomy. Fibrous clots may form in the lumen of the teat at the streak canal subsequent to trauma preventing milk flow.

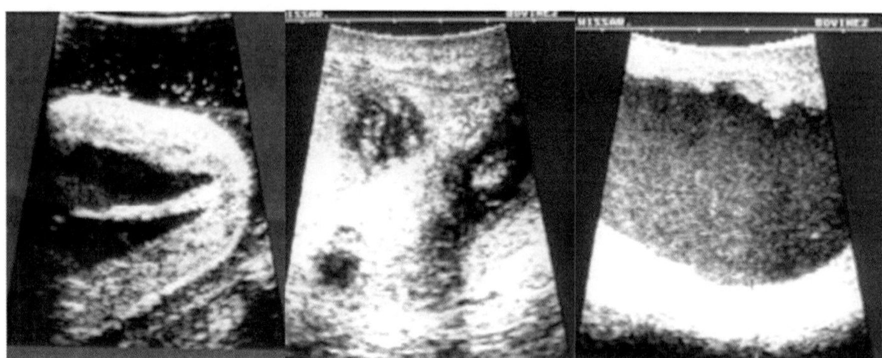

Fig. 87: Ultrasonogram showing teat septum in teat draining from two quarter

Fig 88: Ultrasonogram showing calcification in gland cistern

Fig 89: USG showing accumulation of milk in teat cistern due to teat blockade

21

Congenital Anamolies of New Born Buffalo Calves Ultrasonographic Observations

Congenital anomaly of newborn buffalo calves is quite common in new born calves. Most common encountered are atresia ani, rectovaginal fistula etc. Sometimes, developmental defects interfere with physiological processes viz. in urination and defecation. Therefore, such cases, if not corrected well in time can cause death of the patient. In the present report, few newborn calves were brought to the veterinary clinic with the following abnormalities:

Agenesis of Urogenital System Without Anus

Five newborn buffalo calves were brought to the veterinary clinic with the complain of not passing urine and feces. There was no visible external genitalia and anal opening. At the site of vulva, there was only an impression of vulvar lips and below it, a cord like structure was hanging.

There was an impression of round structure below the skin; in some cases it was outside giving appearance of scrotum which was located in between the two thighs. On ventral side of abdomen, preputial sheath was not present, therefore, it was very difficult to decide about the sex of the calves and site of external preputial-orifice. There was an indication of both sexes, but organs were not developed, therefore it was difficult to decide about the sex of the calves. In some cases, impression of the anal sphincter was visible but there was no visible external anal opening. There was slight deviation of the tail from its base also. Intake of milk was normal. Abdomen was distended bilaterally. Respiratory rate was higher than the normal with slight discomfort .Heart rate was elevated. Blood picture revealed increased total blood cell count with neutrophilia. Radiography of the abdomen showed distended urinary bladder at the floor of pelvis indicating formation of the urogenital system without development of the external opening. Barium meal study revealed marking of gastrointestinal tract and stagnation of the barium up to the rectum.

Ultrasonographic study of the fibrous cord showed presence of lumen in the center surrounded by muscular structure The lumen was observed to be obstructed toward the tip of the fibrous cord in four buffalo calves but in one case there was no visible lumen in the cord All these ultrasonographic findings were suggestive that the cord is a rudimentary penis which could not develop properly. Ultrasonogram further revealed presence of oval hypoechoic structure resembling testicles beneath the skin. There was hypoechoic area also below the tail indicating presence of the vaginal vestibule confirming that all buffalo calves were bisexual.

Externally, it was not possible to decide from where the urine to be taken out because impression of both genitalia (Male and female) were present. Probing of the fibrous cord like structure with a small polyethylene catheter revealed presence of urinary passage except in one calf. From these observations it was inferred that basically all buffalo calves were mixed sex. In four cases, it was easier to approach to the urinary bladder through the fibrous cord to take the urine out. But in fifth calf the urinary passage could not be traced.

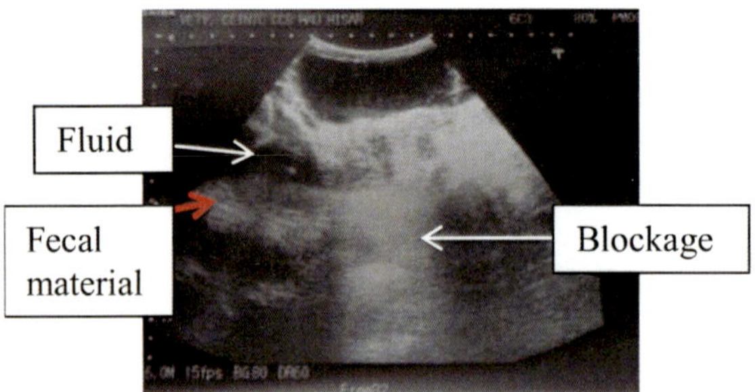

Figure-90 is showing anechoic Urinary bladder toward dorsal side. Below the urinary bladder, rectal pouch filled with anechoic fluid & echoic faecal matter in a calf having antrsia ani.

Fig. 91: is showing striation of rectal wall indicating some infection along with dilated rectal pouch with fluid in a calf.

Congenital Anamolies of Urinary Bladder

In the present report, newborn female buffalo calves were brought to the veterinary clinic with the complain of not passing urine through natural openings In two cases, there was complete retention of urine and in three cases, there was condition of patent urachus ,therefore, the urine was dribbling through the navel area Ultrasonographic study was carried out in all five cases to know the exact cause of retention of urine in first group and to judge the exact attachment of urachus with the urinary bladder in second group so that urachus can be ligated by surgical intervention.

Five newborn female buffalo calves between age group of (2-5days) were brought to the veterinary clinic with the history of problem of urination. In first category of cases, buffalo calves of age of just 2 days were presented with the history of not passing urine immediately after birth. The abdomen was distended bilaterally. External genitalia and anal sphincter appeared to be normal. Examination of genitalia revealed normal vulva and vagina. For further investigation, the calves were sent for radiographic and ultrasonographic examination.

In second category of cases (3 cases between age group of 3-5 days), there was dribbling of urine from the umbilicus and there was no passing of urine from the normal passage. There was slight distension of the abdomen. The defecation of the calves was normal. All calves were sent for ultrasonographic examination to know the site of connection of the urachus with the urinary bladder.

In both categories of cases, the intake of milk was almost normal. Rectal temperature was slightly above the normal. Respiratory rate was higher than the normal with slight discomfort .Heart rate was elevated. Blood picture revealed increased total blood cell count with neutrophillia. Radiography of the abdomen showed distended urinary bladder at the floor of the pelvis.

Ultrasonographic Observations

In first category, the ultra sonograms showed irregular hypo echoic shadow of distended urinary bladder with hyper echoic sediments at its floor .The urinary bladder was lined with hyper echoic membrane present even at the site of urethra. The embryonic membranes lining the urinary bladder were multiple thus preventing the passing of urine from urinary bladder to the urethra. The ultra sonogram of the kidney showed distended hydronephroic kidney. The ultra sonogram of the liver depict normal architecture of the liver

In second category of cases, the ultra sonogram depicts presence of two connections at the periphery of hypo echoic urinary bladder one of urethra and second of urachus. After sonography, it became easy to decide from where the urachus should be ligated.by surgical intervention.

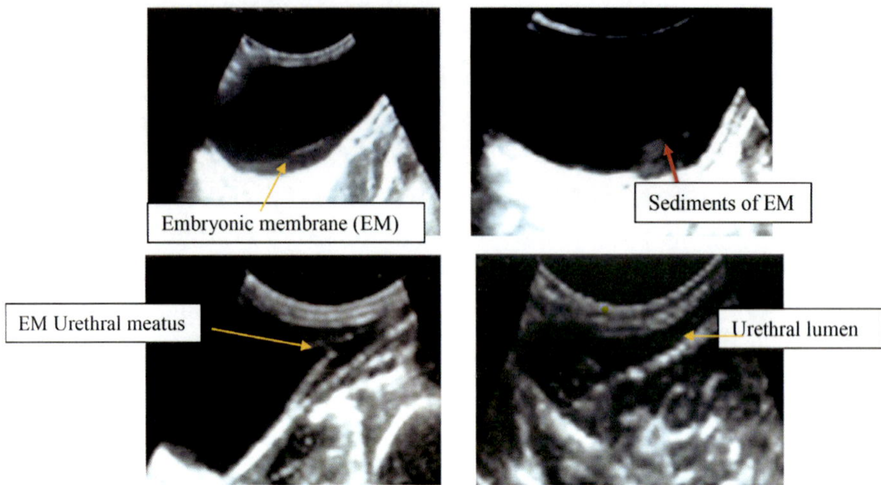

Figure 92-95 is showing lining of the urinary bladder **(top left)** with hyper echoic membrane at the site of opening of urethra **(left bottom)** along with sediments **(top right)** and meatus was occluded by the embryonic membrane at its neck and the urethral lumen was appeared to be patent **(right bottom)**.

22

Three Dimensional and Four Dimensional Ultrasonography in Canine Practice

Ultrasonography has an importance place in human practice and veterinary field especially in canine practice. Since many years two dimensional ultrasound has been used for scanning purposes both in human and animal practice. But the scenario has changed in human practice with development of three dimensional technology and later four dimensional ultrasonography has developed which shows movements of the objects being scanned especially in the case of fetuses. Nowadays it has been common to use these advanced technology in the human practices. Yet these technologies are in the childhood stage in the veterinary field. High cost of 3D/4D ultrasound equipment, operational difficulties in non sedated animals & lack of specific knowledge about the operation of volume scanning hinder the use of this technology in the veterinary field. Better understanding of the different application of these technologies can be made to use in research areas for the better understanding of the abnormalities, post therapeutics as well in the diagnosis of the diseases

Principle of Three Dimensional and Four Dimensional Ultrasonography

Sound waves of frequency higher than human audible frequency usually ranges from 2-13 MHz are used in ultrasonography. Depending upon on depth of tissue from the scanning surfaces frequency is selected. Three dimensional ultrasonography gives three dimensional images of the volume being scanned .it works on the same principle of 2D ultrasonography but ultrasonographic beams are sent different angles to get three dimensional image of object being scanned. Volume information is obtained either directly using 2D array probe or by reconstructing three dimensional volume from a series of continuous or non- continuous 2D images by using software (Fenster and Downey, 2000). An advancement of 3D technology is 4D ultrasound. Four dimensional ultrasound adds the fourth dimension as time and gives a dynamic real-time 3Dultrasonography of the concerned organ with a continuous volume image acquisitioned rendering process. (Khurana and Dahiya, 2004).

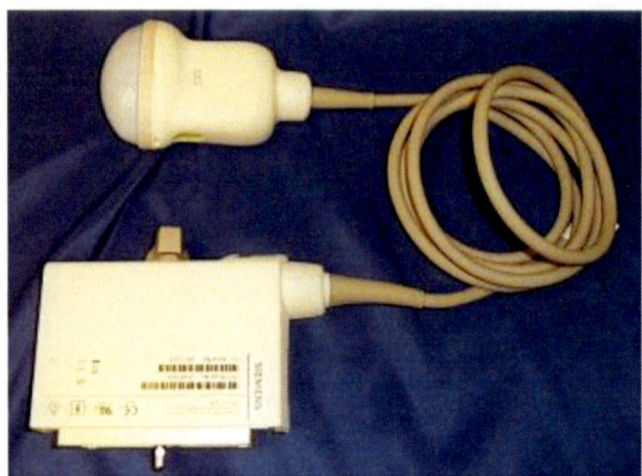

3D/4D volumetric probe

Acquisition of Volume Data

Acquisition and storage of a three- dimensional ultrasound scan can be completed taking less time by an experienced person. But the movement of the animals, probe, panting will affect image quality so better restraining of animals are necessary for the better image quality. Three dimensional ultrasound data can be obtained with mechanical volume probes or by tracking of continuous or non continuous images with conventional two dimensional transducers [Downey *et al*., 2000 Kratochwil *et al*., 1998]. Mechanical probe having rapid oscillation technology is used conventionally in most of the machines. It takes images by rotating the piezoelectric crystals with the help of motor inside the probe and these images are reconstructed in the scanner automatically (Deng and Rodeck, 2004). Freehand systems involve optical or more commonly magnetic field tracking attached to 2D transducer as position sensors. The new generation two dimensional matrix array probes with piezoelectric crystals arranged in two dimensional arrays use electronic scanning to sweep an ultrasound beam over the volume-of-interest to produce 3D images in real time. In 2D matrix array by maintaining the transducer in a fixed position and rotating the volume using the system trackball a 360° rotation and examination of selected structures can be possible Such that an object with its two sides can be visualized in real time with better spatial resolution. (Gonçalves *et al*., 2006)

Image Interpretation

There are several display techniques that can be for analyzes of three-dimensional ultrasound data (Rankin *et al*., 1993). The three basic types of rendering techniques in use are multiplanar imaging, surface rendering, and volume rendering. By multiplanar imaging, the data is visualized in three orthogonal scan planes. While viewing through each two-dimensional scan plane, the corresponding orthogonal planes are displayed immediately. Coronal view can be obtained by rotating the acquired volume (Jurkovic, 2002). In a surface rendering technique, the boundaries of a structure identified are shaded and illuminated giving shape to the surface of the structures. Volume rendering casts rays through volume data and projects the results into a 2D plane. Depending on the algorithm wide spectrum of visual effects can be produced. Minimum- and maximum-intensity projections as well as x-ray (translucency) projections are available which allows visualization of echo poor or echo rich structures. Maximum transparency mode visualizes echo rich structures obtained by scanning with reduced gain. In minimum transparency mode the organs with high transparency are highlighted against more echogenic area. But overlapping of hypo echoic structures can be come as artifacts (Lee *et al.*, 2002). X ray mode displays bony structures by decreasing the ultrasound gain to suppress the muscles and soft tissue imaging surrounding it (Lee, 2003). Tomographic ultrasound imaging cut volume data into number of equal parts and represented in two dimensional representations allowing topographic identification of the volume of interest. The width of each slice and number of the images at a time in the monitor can be adjusted by the operator (Hildebrandt *et al.*, 2009). Four- dimensional images of bones can be obtained by decreasing gain settings only, with no need for cropping. With color Doppler imaging, four-dimensional reconstruction of vascular structures is possible with this new technology. (Goncalves *et al.*, 2006)

Application

The multiplanar presentation and niche mode are quite useful to determine extend of nodules, cysts or tumors and to check whether these are present inside or outside of an organ. Hildebrandt *et al* (2009) used multiplanar presentation mode for the parallel measurements of head (top left), yolk sac diameter (top right) and crown-rump-length (bottom left) in a canine embryo. Niche mode is a cut open view of volume image used to visualize the data inside the volume in a particular plane and can be applied for study of the placental investigation in canine pregnancy. Surface rendering can be used for evaluating the embryonic and foetal integrity in canines (Hildebrandt *et al.*, 2009).The volume measurement of irregular structures are better assessed with 3D and studies can be performed for assessment of growth in order to decide medical or surgical treatment and for the follow up of treatment. The use of this technology in applying color doppler, in guiding needles for different puncture procedures as well in evaluating the fetal heart are currently under close research in human practice and shows future in canine practice also (Purandare C N, 2006). If used along with color Doppler, transparency mode can be used for the study of three dimensional architecture of growing mammary complex in pregnant bitches (Hildebrandt *et al.*, 2009).3D ultrasonography offers a more ample image of anatomical structures and pathological conditions in human practice and also permits observation of the exact spatial relationships(Lees W., 2001 Timor- Tritsch *et al.*, 2002) studies have to undergone more in this area in animal practice with these newer technology.

3D power Doppler ultrasound offers investigation of intratumor vascularization and volume of cervical cancer more easily (Hsu SC *et al.*, 2004). Ultrasonography seems to have many applications for diagnosing of ocular diseases according to anatomical structure of the eye (Janos N., 1997). Caudal ocular ultrasonography is difficult with two-dimensional ultrasound because of the wrong probe positioning and angle. The size and distribution of the tumors are better visible in 3DU especially in the caudal region of the orbit (Romero and Finger 1998) Lesions of the caudal portion of the orbit (e.g. optic nerve atrophy) are better visualized by this technique. A study conducted by Vosough *et al* (2007) 3D ultrasonography of the eye in dogs showed marked advantages in image acquisition for interpretation of all aspects of the ocular structures and suggested that the 3Dultrasound gives useful images for teaching and diagnostic purpose and Lesions of the caudal portion of the orbit (e.g. optic nerve atrophy)can be better visualized by this technique. 3D/4D is widely used to identify fetal anomalies (Tonni *et al.*, 2005, Dyson *et al.*, 2005) in human obstetrics. Increased diagnosis of fetal anomalies has been reported by this technique compared with conventional sonography in human practice (Merz. and Welter, 2005). By offering detailed information about pregnancy status and birth prediction, 3D/4D technology improved the diagnostic confidence. Fetal resorption in animal, ectopic pregnancy can be better studied and can be easily identified. (Hildebrandt *et al.* ,2009) and also aid in better visualization of needle during biopsy(Won.,*et al.*, 2003) .Soft tissue like heart, kidneys, liver, pancreas, gall bladder, eye, and thyroid (Slapa *et al.*, 2011) can also be better visualized. Advantages

There are some features to three dimensional ultrasonography with several universally applicable advantages. In a 3D acquisition, as the sonographer scans the region of interest with a single sweep of volumetric transducer, it will take no longer to acquire multiple image series through the organ of interest, so scan times can be greatly reduced. Data can be stored which can be worked out later also. Virtual planes which cannot be obtained in conventional two dimensional ultrasonographyavailable in 3D ultrasonography. The ability to give better qualitative and quantitative information helps in effective diagnosis. 3D surface visualization is accomplished by this technique improves Detailed anatomical and pathological studies. Precise volume measurement of organs with irregular shape will help in the detection of progression of the diseases and response to the treatment. This may have applications in follicle monitoring in the future (Goncalves *et al.*, 2006).With four dimensional ultrasonography, movement of the internal structures and improved visualization of biopsy devices by more perceptible information on the spatial relationship between the biopsy needle and the target lesion can be obtained(Won.,*et al.*, 2003). 2D matrix array may become an attractive alternative to examine anatomic structures of the adult as well as fetuses, especially fetal hearts enabling view of two different planes of section of the same structure, in real time, without resolution loss (Goncalves *et al.*, 2006). Contrast-enhanced 4D-US can be done for the visualization of the staining of the tumors and the blood flow of surrounding organs three - dimensionally.

Limitation

Although 3D and 4D ultrasonography offers promising future, some limitation has to be overcome by this technology to make it routine in clinical practice. As 3D image acquisition is not fundamentally different from 2D ultrasound, contrast limitations still apply to 3D also. So if there is inadequate amniotic fluid surrounding the fetus, or if the fetus has its face in the posterior position in the uterus, there will be difficulty visualizing structures and the face. False positives can be increased if we ignore this (Maymon *et al* 2000, Fernandez *et al* 2006., Leung *et al* 2005). Breathing and motion artifacts make difficulty to get a clear image with three dimensional ultrasonography. While the artifacts with the breathing movements are not affected by the four dimensional ultrasonography. To date there are no three dimensional / four dimensional ultrasound machines are available only for veterinary practice and hence slow acquisition time and software settings optimized for the human body hinders the resolution of the data obtained. Available Trans abdominal volume probes are bulkier posing some handling difficulty. 4D ultrasound scanner with transvaginal probes are available but are costlier (Hildebrandt *et al.*, 2009). Determination of image orientation becomes difficult to determine as no standardized display convention has emerged till now for the reconstructed images. Rendering of data in 3D using volume or surface rendering techniques introduces an additional layer of potential artifacts and lack of standardized review. Finally, while 3D images are qualitatively more intuitive than traditional 2D sections, data as to the quantitative clinical benefits of 3D ultrasound imaging are limited, particularly compared with the vast volume of literature and expertise available for 2D imaging (Lazebnik. R.S and Desser, T.S 2007).In the 4D ultrasonography with

2D matrix array probe, due to lower transducer frequencies it gives lower resolution than mechanical volumetric transducers and has narrow volume display (Goncalves *et al.*, 2006).

Pregnancy diagnosis with 3D ultrasonography

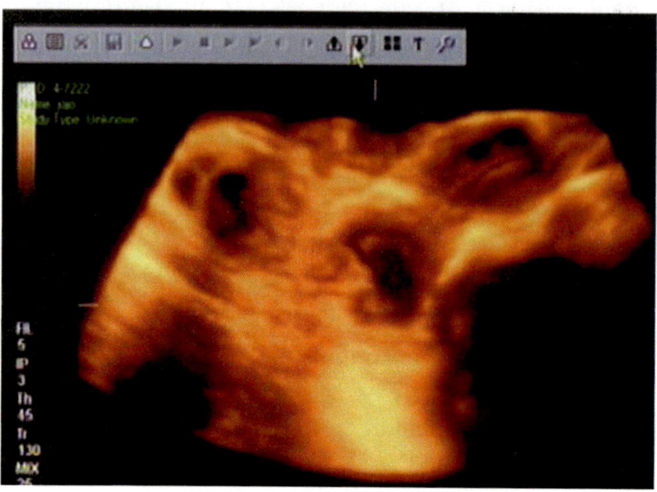

Fig. 96: is showing 3 pups in 3D scan.

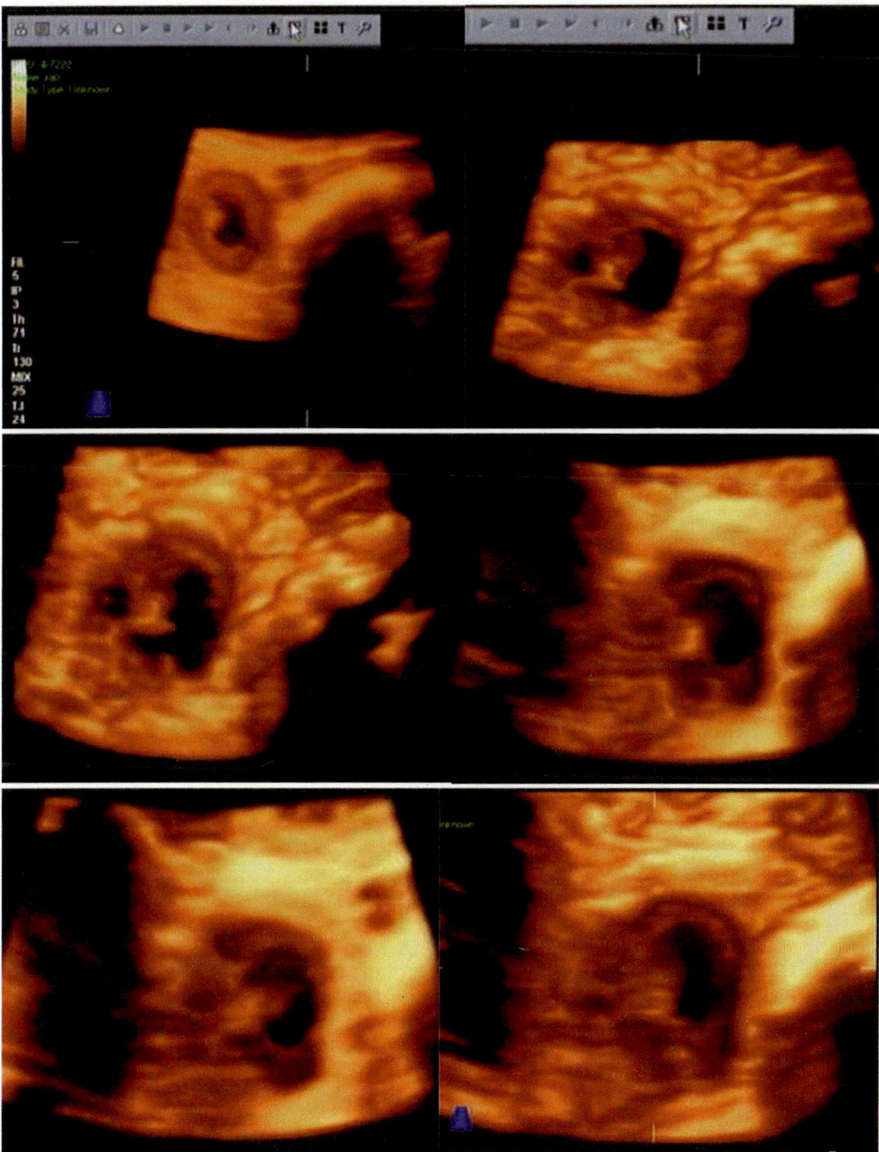

Fig. 97: is collective images of same animal at different angles that were obtained in 3-D ultrasonography, after moving image in three directions.

Conclusion

Three dimensional and four dimensional ultrasonography an emerging modality in human practice and veterinary practice . Still it is considered as an adjunct to conventional 2D ultrasonography (Bega *et al*., 2001). Unlike 2D ultrasound,

3D technology is only in its infancy. higher cost of the machine, limited knowledge concerning the operational skills and difficulty in managing the motional artifacts in the animal, make development of this imaging technique difficult in veterinary practice . Reduced amount of fetal fluid in the small animal compared to human fetus hinders the clarity of the imageswith the machines intented for human practice. Technological refinement and increased standardization of 3D ultrasound evaluation in the coming decades may offer a new opportunity to overcome breathing artifact and motion artifacts during 3D/4D scanning (Hildebrandt *et al.*, 2007).there ismuch research has to be done in veterinary to explore its applicability in the veterinary practice.

23

Principle and Practice of Color Doppler Ultrasonography

In Doppler imaging, the reflections are from circulatory vessels, while in 2D ultrasonic image the reflection is from a tissue(s) or organ(s) In 2D ultrasound strength of reflection in B-mode depends upon density of the organ.On other hand the reflection in Doppler imaging depends upon direction of the circulation. If circulaltion is coming towards transducer, the signal will be strong, while it looses its strength while flowing away from the transducer. As shown in the diagram, the position of transducer on curved surface decides the strength or amplitude of signal. If it is positioned on curve end from where circulation is coming, the signal will be of higher amplitude, but when position is changed to other end of curved surface, where circulation is going down the amplitude of signal will decrease accordingly and is shown in reverse direction.

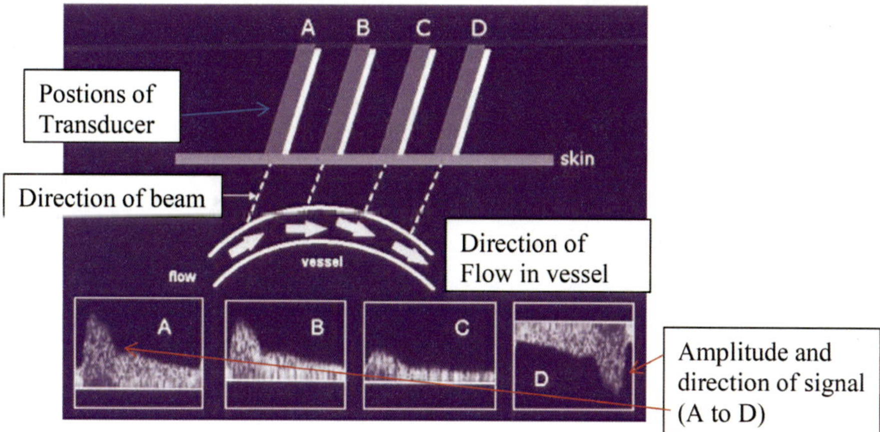

Graphic-4: image is showing four positions of transducer (A,B,C,D) in top panel, while four different magnitude of signals (A,B,C,D) in lower panel. Arrows in curved vessel showed direction of flow. This shows that when blood is flowing towards transducer the signal is shown upward.

Principle of Doppler

Movement of blood in vessels is detected, which is depicted by color in some ultrasound machines. The flow coming towards transducer is shown above baseline, while flow going down i.e. away from transducer is shown below baseline. If position of transducer is changed from where the arterial flow is going away from transducer then, the arterial flow will be shown below baseline and venous flow that is coming towards transducer will be shown above baseline. Some machine shows red and blue color of the flow. The red color shows flow coming towards and blue color shows flow going away from transcer. A beginner ultrasonographer may think that the red color is from artery and blue is from vein. This is not true, this shows direction of flow. Since ultrasonic image of flow in Doppler is formed from series of pulses returning from vessels, the magnitude of pulse changes with variation in flow. A stationary flow will not show any height of magnitude. The magnitude (height) of Doppler pulse changes with power of flow towards or away from transducer and direction of pulse will be changing. In simple words flow incoming towards transducer with power will show higher magnitude on upperside of baseline, while if it is going away, then distance between vessel and transducer will determine downward depiction of magnitude of pulse on downside of baseline. It means time difference of pulse returning from vessel is directly related to magnitude of pulse shown around baseline.

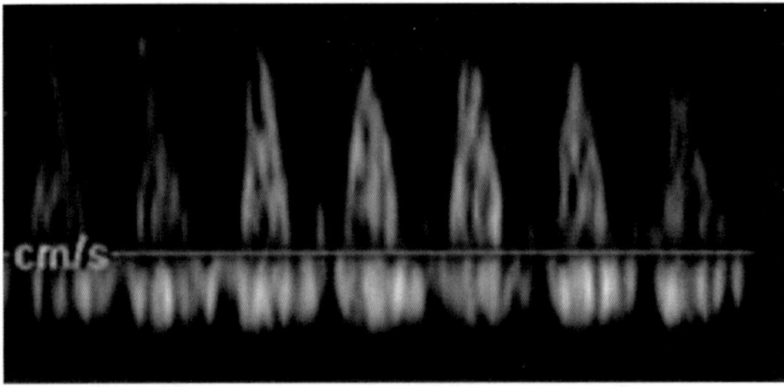

Fig. 98: is showing simultaneous Doppler image of umbilical artery above base line and of umbilical vein below baseline.

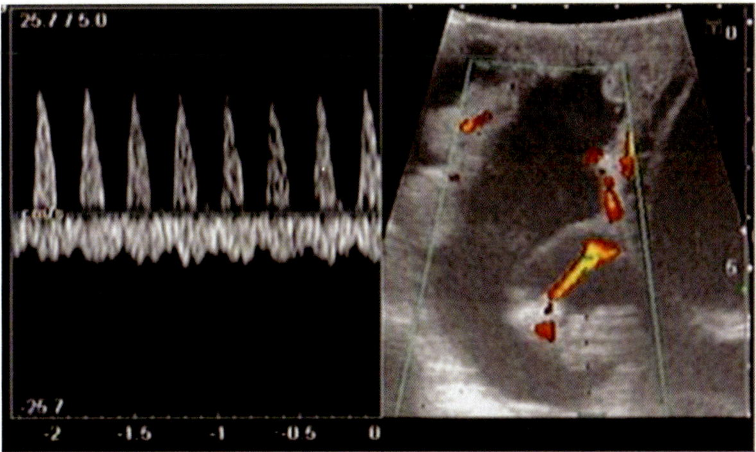

Fig. 99: is showing Colour Doppler image at day 39 of pregnancy in a goat showing pattern of flowing of blood in umblical artery (above base line) and umbilical vein (above baseline). This shows that position of transducer is such that arterial flow is coming towards transducer.

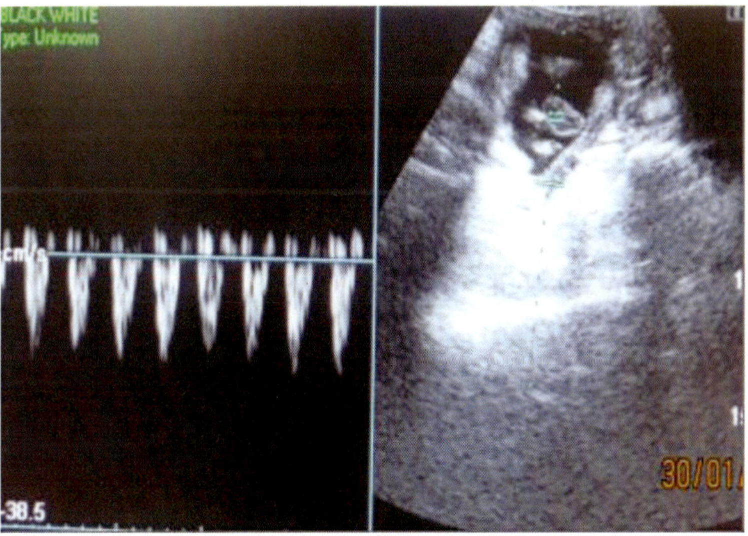

Fig. 100: is showing Colour Doppler image at day 60 of gestation in goat showing pattern of flowing of blood in umblical artery (below baseline) and umbilical vein (above baseline). Only systolic wave is present in artery at this stage of gestation. The presentation of arterial flow below baseline suggesting arterial flow is going away from transducer.

From the above two figures, it is evident that if transducer is fixed at one place without movement, then there is no variation in magnitude of pulses. The magnitude will change if direction of flow or distance of flow changes. In other words, if the flow is perpendicular to the beam of ultrasound (from transducer), then there will be no variation in pulse magnitude. The magnitude of pulse depends upon several factors such as 1) blood velocity 2) frequency used and 3) angle of transducer on flow vessel.

If the pulse magnitude is going too high or seen as tips are missing or tips are shown on bottom side of base line, then the magnitude of pulse can be adjusted either by changing frequency or setting of machine as per guideline given in machine handbook.

Wave pattern in Doppler: There are two types of waves in Doppler imaging 1) Continous wave and 2) Pulsed wave.

In the contiuous wave pattern, tansmission and reception of signals are continuous from top of the organ till final depth (upto which the beam can penetrate). In this case signals obtained from all vessels coming in the way of beam-path are obtained and shown on the screen. This does not specify location of signal (magnitude) within beam path. Such Doppler wave is not good to produce good Color flow image. This is generally seen in small, cheapter machines where the quality of transducer is poor that usually employ continuous wave Doppler on screen. In human sciences, This type of continuous wave pattern has been suggested to use for high velocity of aorta of heart.

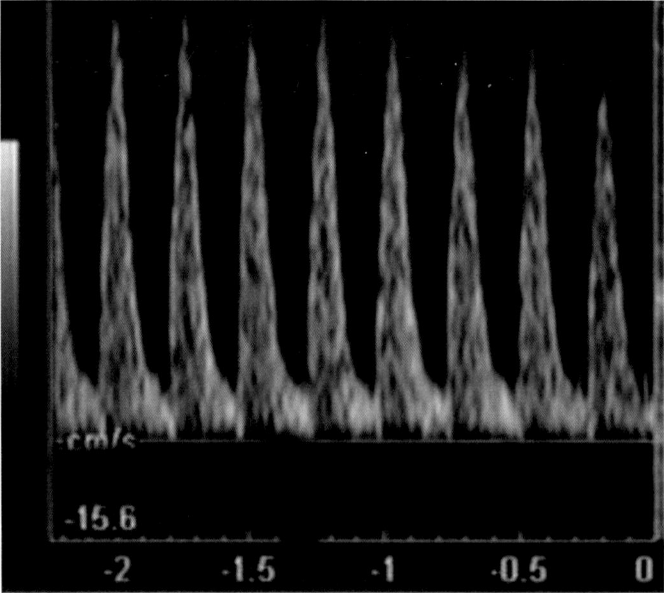

Fig. 101: is showing blood flow in venacava (7th week: Biphasic both systolic and distoic wave present)

Pulse wave Doppler is generally used on Obstetrics to assess supply and exchange of circulation through umbilical artery and umbilical vein. This can also be used to assess supply of pregnant uterus by middle uterine artery. Good transducers employ both B mode imaging and simulateously Doppler wave. Looking at B-mode image location of transducer can be changed to get information about of particular vessel. Therefore, in pulse wave Doppler, to focus on that particular vessel.

Aliasing artifiact: Aliasing occurs due to low frame rate of machine. It also results from wrong use of pulse repeatition frequency (PRF). The low PRF should be used for slow flow velocity. If during scanning, low PRF is on and it strike a high velocity flow. Therefore, to avoid aliasing one sould use low PRF for slow velolcity of flow and high PRF for high velocity flow. When Doppler system is turned on, machine use two pathways. On one side color flow chart is seen and other side flow magnitude is shown as waveform. Aliasing may also happen when a velocity of flow is received at less than the double of the highest frequency coming from flow in the vessel. Infact, the Dopple frequency (DF) is half of RPF. In aliasing the waveform is not completely shown rather its upper tip is shown below baseline as shown in the figure.

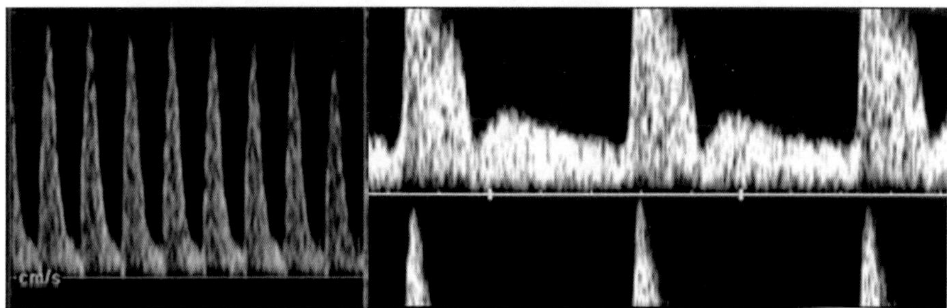

Fig. 102: is showing normal peak on left and aliasing effect on right.

In aliasing, Color pattern is also disturbed. This can be corrected by correcting repeated pulse frequency. Gap between two pulses shoud be sufficient so that one pulse return back to transducer from target organ, before firing of next pulse. If time gap is less then there is overlapping of signals received back from several pulses by the transducer. This overlapping results into aliasing phenomenon. This happen if journey period is long, means when depth of scan is more. As the depth increases the measurable Doppler frequency decreases. Aliasing may also occure, if higher pulse repeatition frequency is used to get Doppler image of low velocity of blood flow. To get image of low velocity, low pulse repeatition frequency should be employed to have sufficient time for getting proper image of low velocity of flow e.g. venous flow. Higher pulse repeatition frequency should be used high velocity of flow. Mismatch of pulse repeatition frequency and flow velocity leads to aliasing.

Color Flow Modes in Doppler Ultrasound

A region needs to be selected to get color flow imaging in Doppler system. If the selected region is large, then amount of information about color imaging of the area does not yield specific information. The selected region should be as small as possible. There are two modes of Doppler ultrasound flow 1) Color Flow Imaging Doppler (CFID) and 2) Spectral Doppler or pulsed wave Doppler (PWD).

- Color flow imaging (CFI) may generate overall view of flow in different colors of a selected region, but information is general and of limited use. However, this can be useful to identify vessels and their direction of flow in the selected region. It gives overall view of flow in the area. However, the resolution of color flow is poor. It can indicate whether flow is of high velocity or low velocity. Therefore, select a small box covering area of interest.

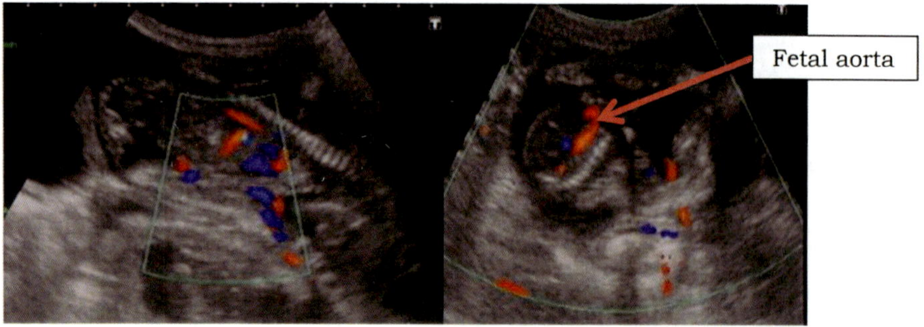

Fig. 103: is showing Color Flow imaging of fetal circulation in a bitch. The vessels can be identified from color flow imaging. Note color flow image of fetal aorta in right image. Note color flow image of fetal aorta in right image.

- CFID produces color image of flow and shown in B mode image. Since color generation depends upon movement of blood flow. Therefore, one should be careful, not to shake the transducer, otherwise several color artifacts will be generated on screen. Transducer elements have to do dual functions for CFI; switching between B-mode and Color flow. It has been shown that the pulses used by transducer for CFI are several folds longer than pulses used for B-mode imaging. Color generated may depend upon direction and magnitude of flow. Flow coming towards transducer is shown in red, while flow going away is shown in blue. Color pattern can be reversed by changing setting of machine.

- Generate of good quality image in CFID depends upon some factor. Proper setting and usage of these factors can improve quality of color flow image. One should take care of followings for good quality CFI: i) Optimal setting of Gain of machine ii) Proper positioning of transducer and focusing on vessel(s) of interest to detect velocity of blood flow iii) Proper adjustment of Pulse repeatition frequency or scale to cover the flow area iv) proper size of box over region of interest. Smaller the size of box may have better color resolution and frame rate.

On other hand, Pulsed wave Doppler (Spectral Doppler) generates specific information of flow in a vessel at that site. This is good to assess blood flow at one place in a specific vessel. It provides view of distribution of flow with good resolution of flow. This is used to calculate velocity of flow and other indices of flow. To obtain clear signal in pulse wave Doppler (PWD) one should be knowledgeable to use proper power and gain of machine. PWD uses more intensity of power than B-mode ultrasound.

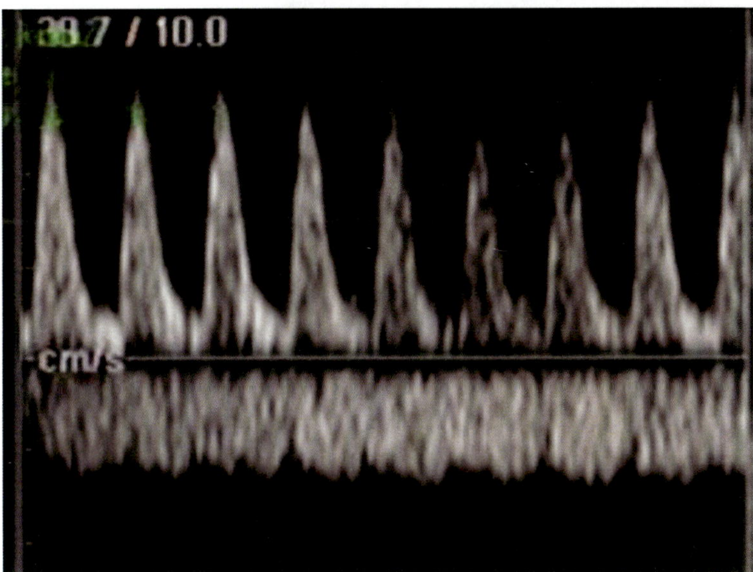

Fig. 104: is showing pulse wave Doppler of umbilical artery above and vein below of fetus inutero in a bitch.

To get better image with PWD, low PRF should be used for low velocity of blood flow, however, if aliasing is seen on screen, then higher PRF should be changed to. In PWD to measure flow in a vessel a proper gate size should be selected. The gate size should be as such that can cover just the vessel. There are setting in good ultrasound machine that provide a knob to change gate size. Large gate may select two neighbouring vessels leading to overlapping of signal. Therefore, proper gate size focusing on one vessel at a time as per size of vessel. Angular focusing of transducer on a blood vessel produces better image than right angle focusing on that vessel.

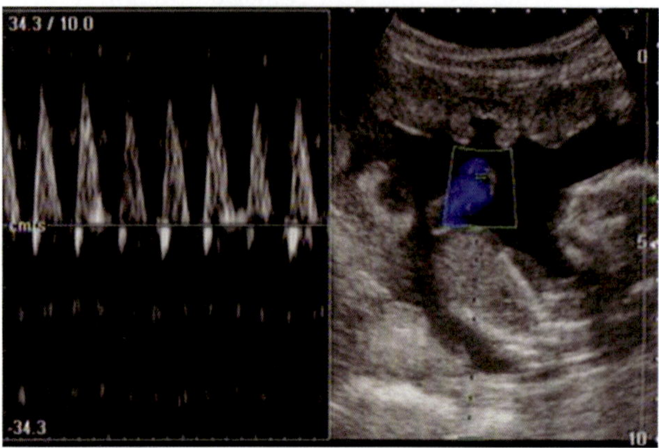

Fig. 105: is showing small box selection on umbilical vessels in a in utero pup

When both Color flow imaging and Pulse Wave Doppler are used, one can get better information.

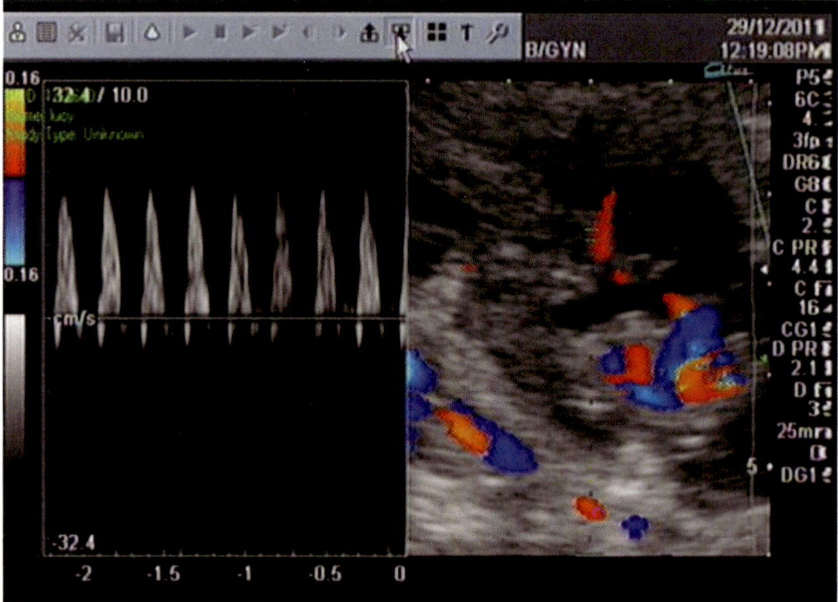

Fig. 106: is showing color flow image on write and waveform on left from fetus and uterus of pregnant bitch

In combined imaging the performance of each component is decreased, because all three components of transducer are working simultaneously i.e. B-mode, color flow and pulsed wave Doppler. In this state, there is more chance of aliasing due to reduction in all three segments i.e. pulse repeatition frequency, frame rate, size of color flow box. Therefore five points are stressed for better imaging with PWD (Spectral Doppler); i) focus the PWD-cursor on particular vessel with proper gate size ii) Gain and power should be adjusted to get clear image iii) position the transducer angular to blood vessel avoiding 90 degree focusing iv) Base line and scale of RPF sould be adjusted to avoid aliasing v) Sample volume selection through proper gate size adjustment.

Use of Power Doppler: when flow is low or minimum, then to increase sensitivity of detection of flow the Power Doppler is used. Power Doppler is also referred as energy Doppler. Some ultrasound expert also call it Doppler angiography or amplitude Doppler. It does not indicate direction of flow. Only magnitude of color flow output is displayed. It increases sensitivity of low flow and is used in conjuction with other two types of Doppler (color flow imaging and pulse wave Doppler).

Information Drawn from Flow Waveform

Shape of Flow Waveform: From visual inspection of waveform shape can easily provide valuable information e.g. as shown in figure below the flow in umbilical artery of developing pup in utero has only systolic wave, while diastolic wave is missing. As the pregnancy advances, the diastolic wave appear at later stage of pregnancy.

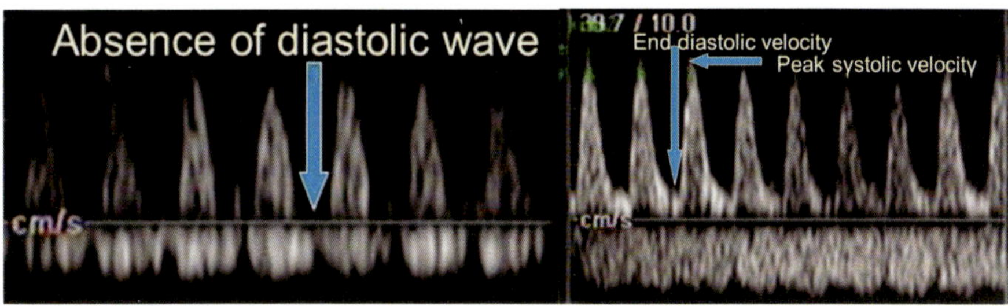

Fig. 107: is showing only systolic velocity in umbilical artery on left in early pregnancy while appearance of end diastolic velocity in mid pregnancy. Other common indices used are Resistance Index (RI), Pulsatality index (PI) and systolic/diastolic ratio (S/D).

1. $$\text{Resistance Index} = \frac{\text{Systolic peak velocity} - \text{End Diastolic velocity}}{\text{Systolic peak Velocity}}$$

$$RI = \frac{(SPV - EDV)}{SPV}$$

2. $$\text{Pulsatility Index} = \frac{\text{Systolic peak velocity} - \text{End Diastolic velocity}}{\text{Time Average Maximum Velocity}}$$

$$PI = \frac{(SPV - EDV)}{TAMX}$$

Timed Average Maximum Velocity (TAMX) is also known as Amplitude weighted frequency.

Note: Systolic peak velocity can be measured from baseline to tip of systolic peak. Similarly End Diastolic Velocity can be measured from base to its height. Timed average Maximum Velocity can be measured from number of cardiac cycles and average of heights of these peaks. In fact, when blood is flowing it has different velocity at different time points. Therefore, moving transducer over vessel to assess velocity at different time points and taking average may lead to Timed Average Maximum Velocity per unit time e.g. 42 cm/sec.

24

Diagnostic Imaging in Common Surgical Conditions of Buffaloes

Diagnostic techniques are routinely used for diagnosing different surgical conditions. The most frequently used imaging techniques in veterinary practice are radiography and ultrasonography. Radiography and ultrasonography are not suitable for each organ in large animals. In small animals most organs are suited to ultrasound images except bones and gascontaining structures such as lungs etc. Bone and gas reflect ultrasound and do not allow the sound waves to form image therefore radiography gives better diagnostic images for these organs. The liver, spleen, kidneys, prostate, gravid uterus and fluid containing organs like urinary bladder, gall bladder and major blood vessels can be better imaged with ultrasound. In buffaloes in surgical section important clinical conditions relate to urinary and digestive systems, mammary gland and perineal area.

1. Urinary System: In bovine, the cases of obstructive urolithiasis are commonly observed. These types of cases are more common in bullocks and in buffalo calves. In these cases radiography is not so useful. Sonography can be tried to take the image of ruptured urethra in bullock and can be compared with radiography. Due to ruptured urethra there will be accumulation of urine in the s/c tissue of prepuce. Due to fluid density the image will be radioopaque and will appear white and no detail of the organ can be seen. The ultrasonogram (USG) will show images of glans penis and its coverings. Different diameters of urethra and a point of rupture just behind to glans penis can be make out. Due to obstruction, The B.calves either will pass drop wise urine or there will be complete stoppage of the urine. These cases can also be easily diagnosed by ultrasonography.

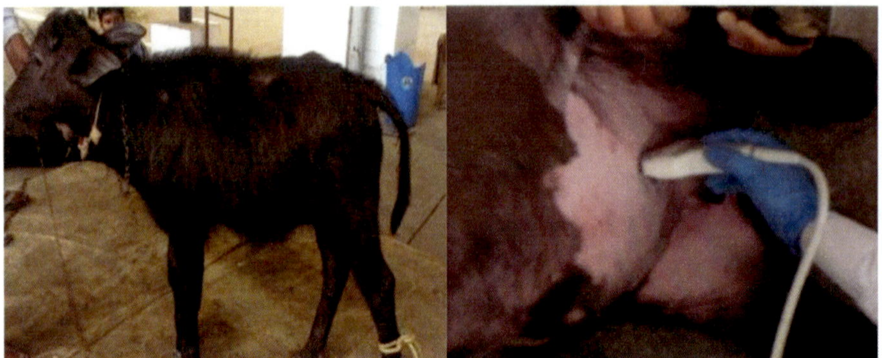

Photographs- 26-27: are showing a calf having urine obstruction **(left)** and scanning of the calf **(right)**.

Due to obstruction of urethra by concretions, there will be accumulation of urine in urinary bladder (U.B.) resulting in its dilatation. Due to accumulation of urine for a longer period, the density of urine gets also increased and on USG, these changes can be noticed. Scattered echogenic particles can also be seen. Other cause of retension of urine observed was inflammation of urethra resulting in distension of urinary bladder. 2D & 3D USG images of distended urinary bladder can be taken and compared. In both dimensions, changes in density of urine and wall of U.B. can be observed and equally appreciable.

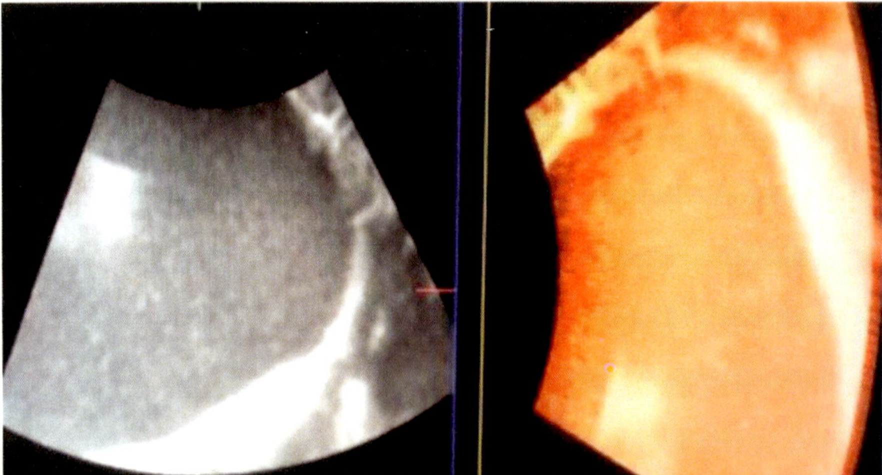

Fig. 108 is 3D scan of urinary bladder with cystitis and abnormal urine in a buffalo-calf.

If the Buffalo-calf is not treated well in time for retension of urine than after a gap of 72 hours, the urinary bladder will get ruptured. Due to rupture of urinary bladder there will be accumulation of urine in the abdominal cavity resulting in bilateral distension of the abdomen. This condition can be successfully diagnosed by ultrasonography.In cases of rupture of urinary bladder, there will be complete disruption of hyperechoic wall of the urinary bladder. The echogenicities of urine in urinary bladder and in abdominal cavity will be same. Moreover in ruptured urinary bladder, the wall of urinary bladder can not be demarcated.3-D image of ruptured urinary bladder can be useful to find out the point of rupture of bladder wall. The accumulation of urine in abdominal cavity can also be visualized as anechoic image with floating of hyperechoic intestinal loops. The rupture of bladder wall can be more clearly seen in 3-D USG. These cases can also be successfully managed by tube cystotomy. It can be concluded that different clinical conditions of urinary system cystitis, cystorrhaxis and complications of tube cystotomy in B.calves can be studied by USG. 3-D USG was observed to be better than 2-D USG to delineate point of rupture of wall of U.B.

2. Digestive system: Themost important surgical condition of the digestive system is foreign body syndrome. The presence of foreign body in reticulum can lead to complications such as traumatic reticuloperitonitis (TRP), reticular hernia and peri-reticular pathology.The incidence of TRP in buffaloes is higher than cattle in India. In ruminants, peritonitis that develops due to foreign bodies is mostly localized. The diagnosis can be made on basis of history and clinical signs. The blood examination will show Neutrophilia with shift to left. Plain radiograph of the reticular area will confirm the presence of foreign bodies. Therefore, radiography is observed to be very successful in diagnosis of lesions of reticulum and diaphragm. The reticular hernia is the passage of abdominal viscera into the thoracic cavity through the congenital or acquired opening in the diaphragm. Generally it is the reticulum which herniates in to the thorax but sometimes other organs such as liver, spleen, omasum, abomasum, and loop of intestine also herniate. Plain or contrast radiograph of the reticular area can confirm the presence of reticular or diaphragmatic hernia in buffaloes.

Recently ultrasonography has also been tried for lesions of reticulum and diaphragm in buffaloes e.g. traumatic reticulitis, reticular hernia, peri-reticular abscess and fistula. The ultrasographic image of metallic foreign body in traumatic reticulitis is observed to be hazy as compared to radiography. In Reticular hernia, the reticulumcan be seen as irregular structure with heterogenous contents having thick echogenic wall herniated in to the thoracic cavity. The reticular abscess can be depicted as hypoechogenic cavity surrounded by echogenic capsule with heterogenous contents partitioned by echogenic septa. The reticular fistula can be delineated as thick walled echogenic tubular structure having anechoic lumen.

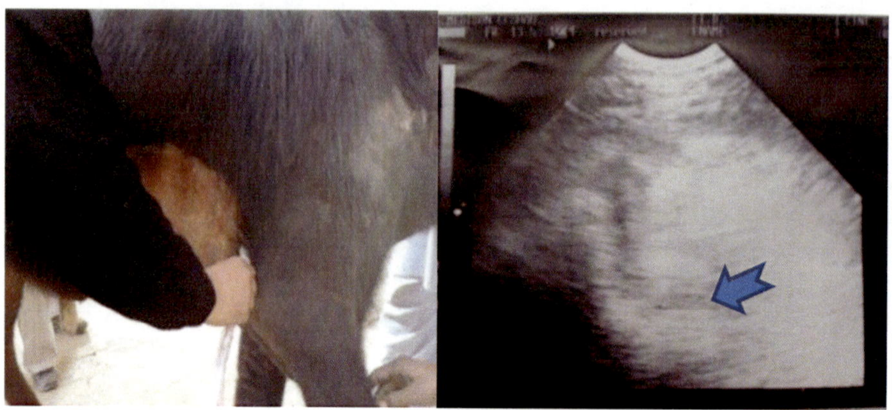

Photograph-28 and Figure-109 is showing site of scanning of reticulum **(left)** and peri-reticular abcess **(arrow)**.

The migration of foreign body from the reticulum can also result into pathology of the surrounding structures i.e. peri-reticular pathology. These perireticular pathologies can be peri-reticular abscess and reticular fistula etc.

These pathologies can also be diagnosed by both radiography and ultrasonography. Radiograph will show Contracted and lifted reticulum with radiolucent zone indicating reticular abscess. The USG gives better image than radiograph so far diagnosis of the reticular fistula is concerned. The cases of traumatic reticulitis, reticular hernia and reticular abscess can be treated by surgical intervention.

3. Mammay gland: Diseases of the mammary gland (M.G.)in buffaloes cause direct economic loss to the farmers. Mastitis is most common disease affecting adult lactating buffaloes. In buffalo heifers the development of the mammary gland can be hindered by mammary tumours. In initial phases, the mammary tumour usually affects one quarter but it can affect more than one quarter if not treated well in time. The milking ability of these quarters can only be known after parturition. The age of these buffalo heifers was more than one years and these B.heifers were in different stages of pregnancy. In these buffalo heifers only left hind quarter was involved but In two cases, two quarters were involved. The affected quarter was swollen & hard to touch. The remaining quarters were soft without any swelling. Initially the swelling was small nodular in size but later on it assumed big size. The radiograhy is not useful for M.G. tumours because no internal details are possible therefore affected swollen quarters were subjected for ultrasonography. USG picture of these hard tumours showed hyperechoic capsule encircling the tumour. Echogenic wavy striations present through out the body of the tumour. These echogenic striations are of fibrous tissue and are indicative of benign tumour. USG picture of these soft tumours showed multiple lobes & septa. Lobes were filled with anechoic fluid. These multiple lobe & septa are indicative of malignant tumour.

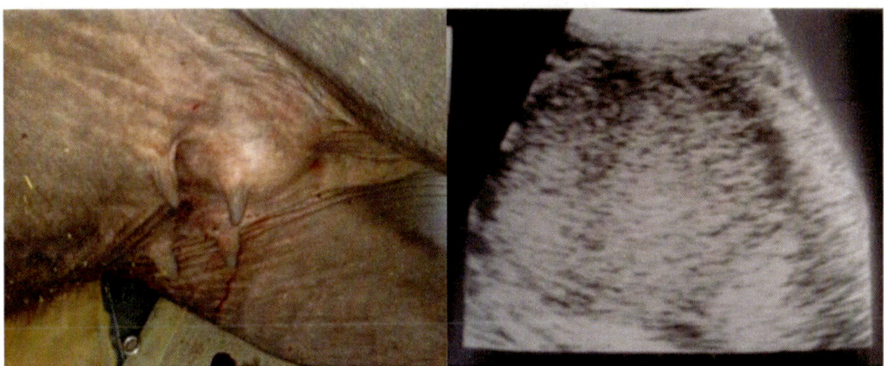

Phtograph-29: (left) and Figure-110 (right) are showing mammary tumor and its ultrasonic image in a buffalo.

These tumours were removed by surgical intervention. In first type, the removed mammary tumours were round, capsulated and hard to touch. In second type, the removed mammary tumours were diffuse and soft to touch. The histopathological examination showed five cases were of fibro-adenoma and three cases were of adeno-carcinoma. After removal of the mammary tumour from affected quarters, remaining quarters showed full functional ability after parturition.

4. Perineal area: Diagnostic Imaging can be used successfully to image perineal pathology. The radiography is not so useful in perineal area as compared to ultrasonography. The perineal pathology in buffaloes can be due to hernia, accumulation of pus or fluid in the perineal area. The perineal swelling was due to constant straining over a month in first type of cases. In second type, the swelling appeared after an accident. In third type,there was gradual increase in size of the perineal swelling without any history of trauma or straining. In 1st group, the swelling was big and toward latero-ventral side of the vulvar lips. In 2nd group, the swelling was very big and extended from latero-ventral side of vulva up to the anus in the buffaloes which met an accident. In 3rd group, the swelling was present just below the vulvar lips and was big in size. The swelling was constantly increasing in size. The haematological values of these animals showed decreased value of haemoglobin, increased value of total leucocytic count, specially neutrophils and increased value of packed cell volume. The ultrasonographic study of these perineal swelling showed following findings. In first group of perineal hernia, the size of the hernial ring was large in size. The hernial content observed were intestine in three buffaloes and urinary bladder in one buffalo. In 2nd group, there was no hernial ring and there was only fluid accumulation. In 3rd group also there was no hernial ring and the content of the swelling was purulent material.

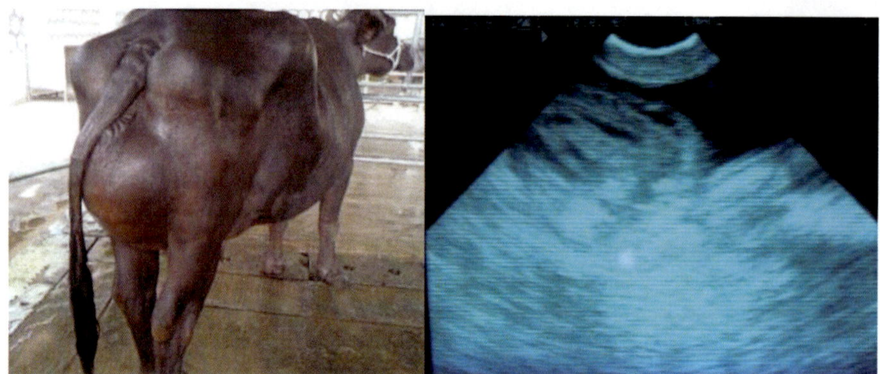

Photograph-30: and Figure-111 are showing a buffalo having perineal hernia **(left)** and its USG image **(right)**

For treatment, the buffaloes were controlled in standing position in *Travis* following sedation with *xylazine.* The caudal block was achieved between sacro-coccygeal joint after injection of 2% lignocaine Hcl. In 1st group of perineal hernia, the rectal contents were evacuated by hand and a guage plug was placed in the anal opening. A slightly curved incision about six inches long was given toward lateral side of the anus to expose the hernial contents. The intestinal loops and the urinary bladderwere pushed back to its normal position. The defect in pelvic diaphragm was repaired by placing sutures in two layers. In *2nd group,* thedead space was obliterated by suturing muscles with surrounding tissue using *catgut* no. three as suture material. In 3rd group, the purulent material was drained off. Postoperatively, the buffaloes were maintained on full course of antibiotics, analgesics & anti-inflammatory drugs, B-complex and fluids. The buffaloes were given liquid paraffin for four days and maintained on green fodder only for two weeks thereafter routine diet was given. The surgical wounds were dressed daily till removal of skin sutures after 14 days.

Radiography gives better diagnostic images for hard structures such as bone and urinary calculi etc. Ultrasonography is more suited for fluid containing soft tissue densities. Therefore fluid containing organs like urinary bladder, reticular abscess, fistulae can be better imaged with ultrasound. Ultrasography is better diagnostic technique as compared to radiography to diagnose clinical conditions of mammary gland and perineal area. For better interpretation, both techniques should be used simultaneously.

25

Ultrasonogaphy of Gastrointestinal Tract In Canine

Ultrasonography of the GI tract presents problems and challenges for the sonographer. Sonography of the GI viscera was once throught to be impossible because of artifacts created by gas within the lumen but with the technical advancement in the ultrasonography now the problem has been solved. If the animal is kept of feed and adlib water is given to the dog with I/v fluid than ultrasonography becomes helpful in the diagnosis of gastrointestinal tract disorders. Other conditions such as esophageal carcinomas can also been diagnosed by ultrasonographic endoscopy. in human beings.

The normal ultrasonography of the liver, kidney, stomach & spleen in dogs has been standardized and its use in evaluation of pathological lesions has been reported but ultrasonographic evaluation of oesophagus & intestine, both in normal animals and diseased conditions is yet to be established. Therefore normal ultrasonography of the oesophagus & small intestine in dogs and the importance of diagnostic capability of ultrasonography in diagnosis of pathology of these organs have been compiled.

a) Normal ultrasonography of oesophagus and small intestine Ultrasonographic appearance of oesophagus:-

20th day

The lumen of the oesophagus was anechoic and wall of the oesophagus was hyperechoic in appearance. Just above the oesophagus jugular vein was present and the lumen of it was also anechoic. The wall of jugular vein was hyperechoic in appearance. Just beneath the oesophagus incomplete tracheal rings appears on the sonogram which were hyperechoic in appearance. Oesophagus appears to be uniform in size from pharynx to the thoracic inlet.

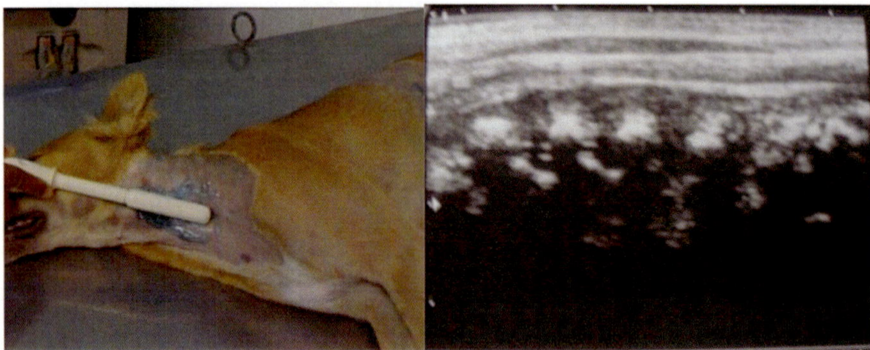

Photograph-31 and Figure-112 is showing scanning position of oesophagus **(left)** and its USG image **(right)** in a dog.

1st month

There was increase in the lumen of the oesophagus but the wall thickness remained the same. The total diameter of oesophagus also gets increased.

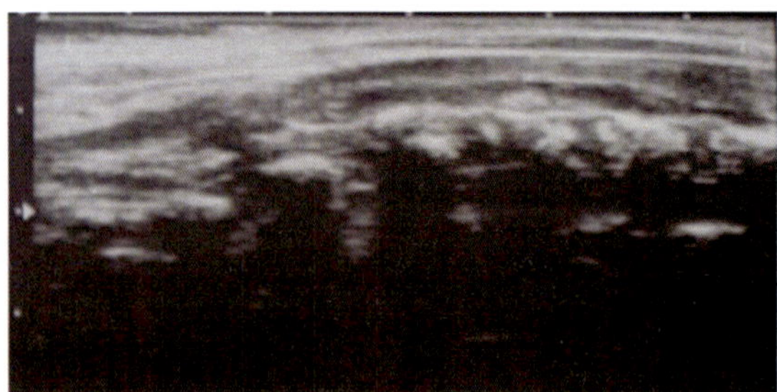

Fig. 113: is showing image of oesophagus at 1st month

2nd month

There was increase in the size of the lumen of the oesophagus but there was no change in the wall thickness.

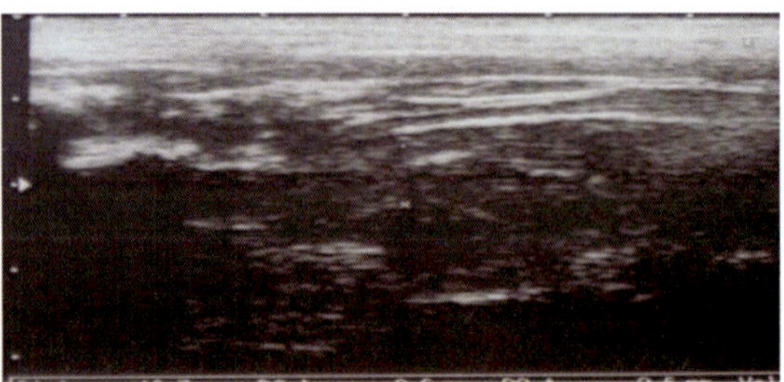

Fig. 114: is showing image of oesophagus at 2nd month

5th month

At this age group, the lumen, wall thickness and total diameter of the oesophagus was observed to be increased. Changes in total diameter were almost same. At this stage, the main finding was appearance of distinct layers of the oesophagus.

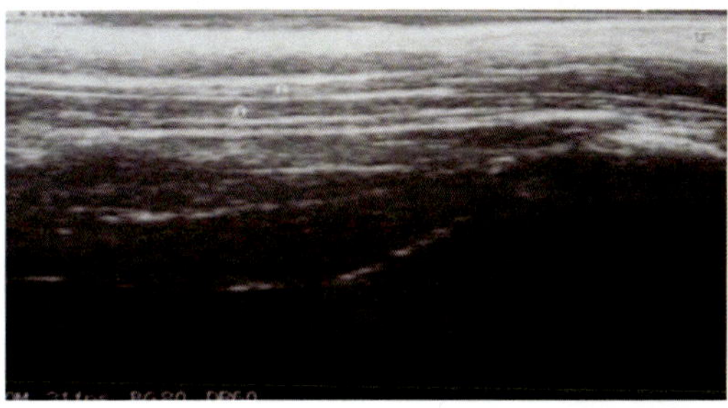

Fig. 115: is showing image of oesophagus at 5th month

6th month

At this age group, the lumen, wall thickness and total diameter of oesophagus was found to be increased but the observation was same as that of 5th month. The total diameter of oesophagus was observed to be more.

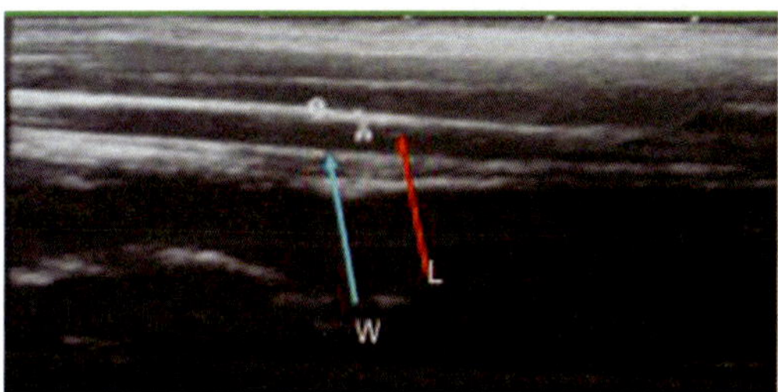

Fig. 116: is showing image of oesophagus at 6th month

Ultrasonographic Appearance of Small Intestine

20 days

The sonogram at this stage, gave clear image of stomach, pyloric end of the stomach and loops of the small intestine. The image of stomach was anechoic with presence of rughae as hyperechoic structure in the centre of the stomach. Opening of the stomach in to the intestine i.e. pylorus can be seen as thick walled hyperechoic sphincter surrounding the anechoic lumen. After pylorus the lumen of the duodenum can be seen as anechoic structure surrounded by hyper echoic wall.

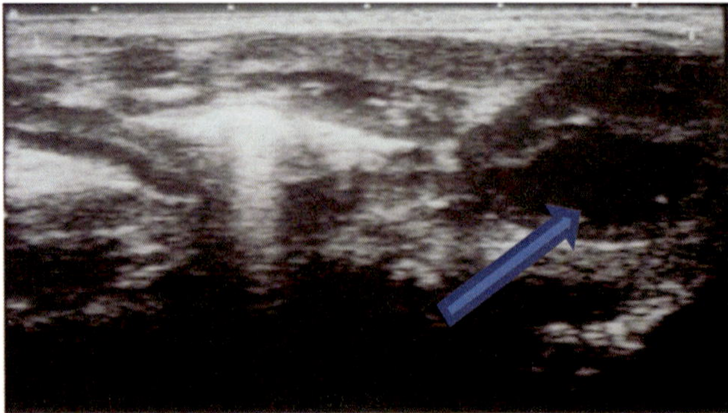

Fig. 117: is showing image of stomach at day 20 in a pup.

1st month

There was increase in the lumen diameter, wall thickness and total diameter of the stomach and intestine.

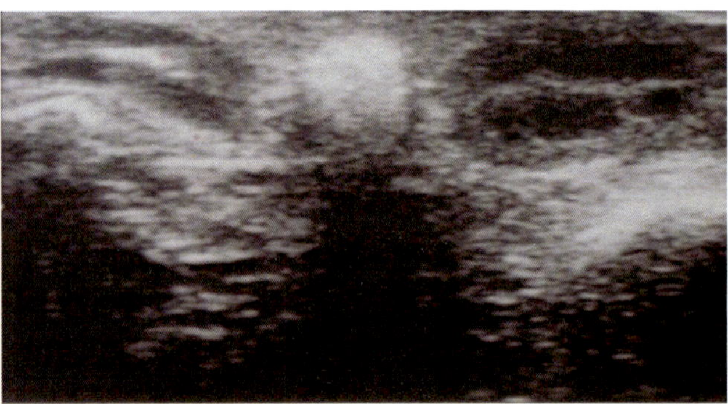

Fig. 118: is showing image of stomach of a pup of day 30.

2nd month

- At 2nd month of age, all the layers of the small intestine starting from the lumen i.e.
- tunica mucosa
- tunica submucosa
- tunica muscularis
- tunica serosa

were found to be distinct. The size of lumen, wall thickness and total diameter was observed to be increased.

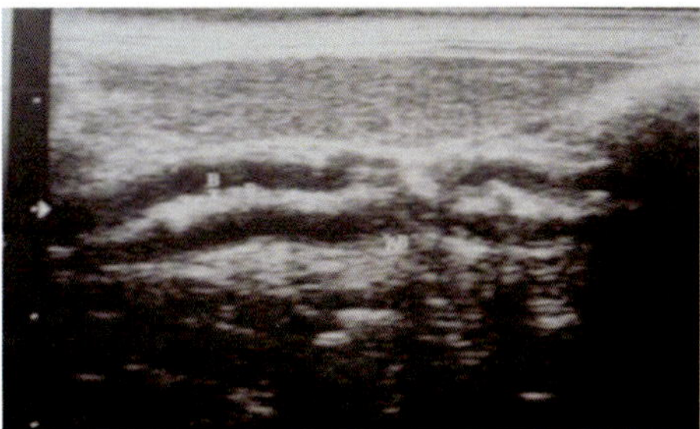

Fig. 119: is showing image of intestine at 2nd month of age of a pup.

5th month

At this age, there were many cross sections of the small intestine indicating formation of many loops of small intestine. These cross sections produced a target sign or bull-eye appearance.The lumen diameter, wall thickness and total diameter was observed to be increased.

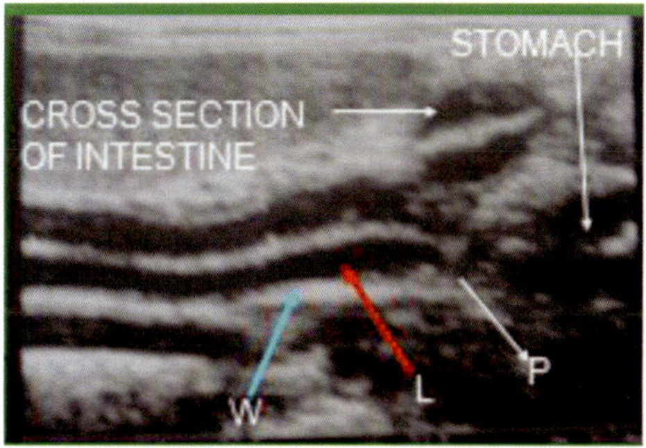

Fig. 120: is showing image of intestine at 5th month of age of a pup.

6th month

At 6th month of age, total changes in intestine were observed to be increased. The image of fully developed intestine was found at this stage.

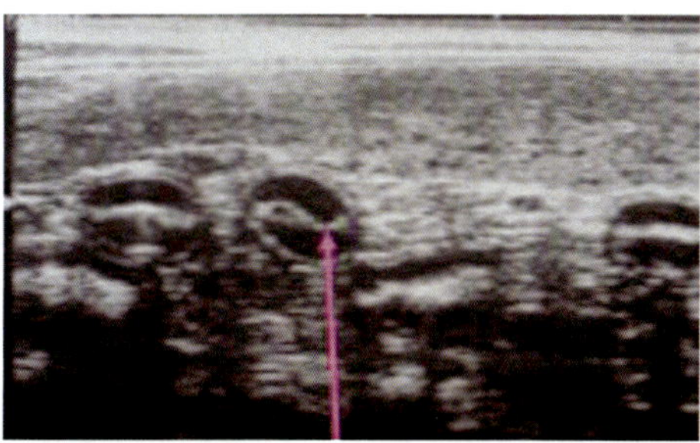

Fig. 121: is showing image of intestine at 6th month of age of a pup.

b) Clinical Conditions of Oesophagus & Intestine

Conditions of the oesophagus & intestine studied by ultrasonography in dogs are as follows:

1. Mega-oesophagus
2. Oesophageal obstruction
3. Ulcerative enteritis
4. Pica
5. Foreign bodies
6. Atony of intestine
7. Intussusception
8. Retained meconium

1. Mega- oesophagus

History: A 20 days old pup was admitted with the history of swelling on the ventral side of the neck. There was difficulty in taking of food and water since 7-8 days. The pup was showing sign of vomition after intake of food and water. There was coughing for the last 7-8 days.

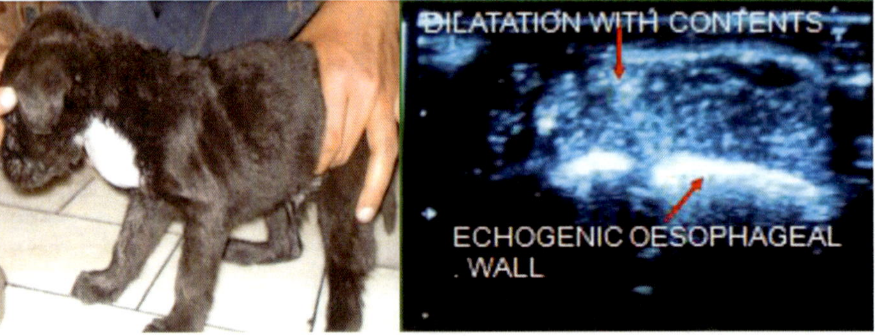

Photograph-32: and Figure-122 are showing case **(left)** and image of Mega- oesophagus **(right)** in a dog.

Ultrasonographic Interpretation

On ultrasonography, there was appearance of dilated oesophagus, which was hypoechoic. The contents of the lumen were echogenic indicating accumulation of air and the wall of the oesophagus was hyperechoic in appearance.

2. Oesophageal Obstruction

History: A dog was admitted with the history of difficulty in taking food and water since 5 days. There was vomition immediately after food and water intake. There was coughing also for the last 5 days.

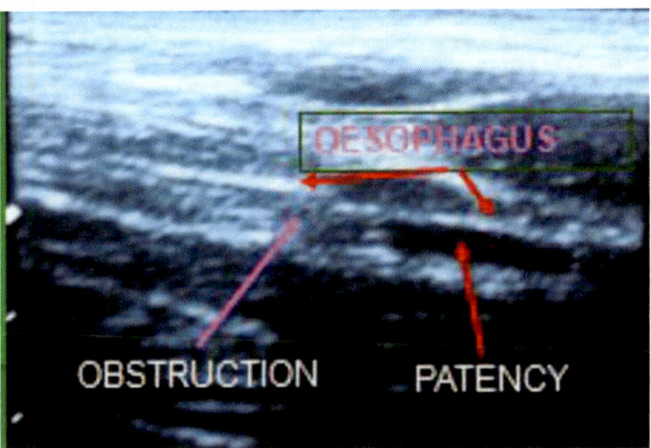

Fig. 123: is showing image of oesophageal obstruction in a dog.

Lumen of the oesophagus was anechoic towards cranial side indicating normal patency with accumulation of the fluid and it was hypoechoic at the site of obstruction indicating blockade caudally. The walls of the oesophagus were hyperechoic in appearance revealing inflammatory changes.

3. Ulcerative Enteritis (parvo-virus infection)

History: Dogs were admitted with the history of dysentery since 10-15 days. Dogs were off feed for the last 8-10 days. Dogs were vomiting since 7-9 days. The dogs were not responding to the medicinal treatment.

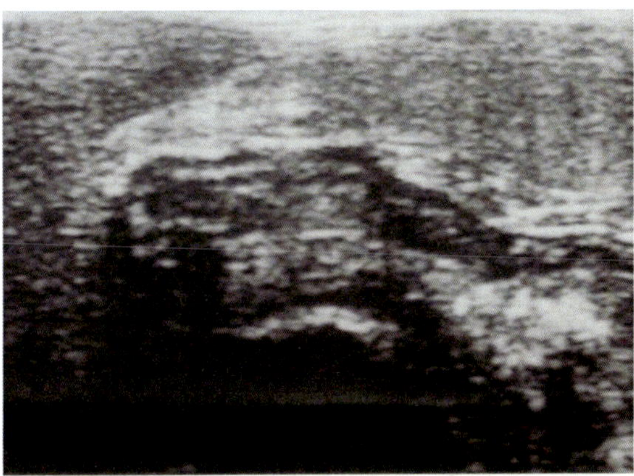

Fig. 124: is showing image of ulcerative enteritis in a parvo case of a dog.

On ultrasonography, there was sloughing of hyperechoic mucosa from the wall of the Intestine. The lumen of the intestine was having anechoic liquid fecal contents and the echogenic accumulated gases. There was breakage in the continuity of the intestinal mucosa at many places.

4. PICA

History: Two dogs were admitted with the history of eating of sand since one month. The dogs were not dewormed since birth.

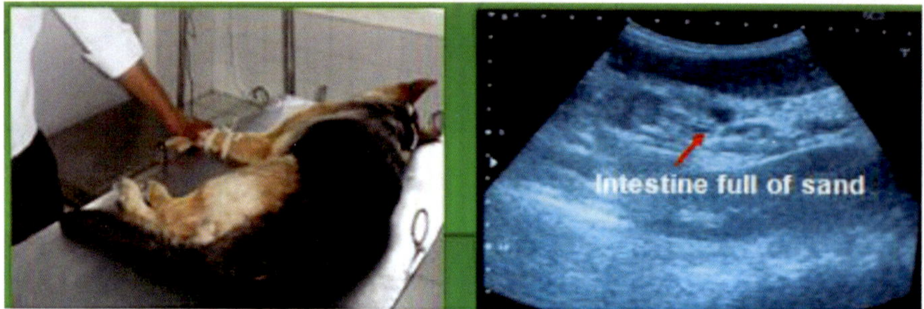

Photograph-33: and Figure 125 is showing a dog (left) and image of intestine (right) containing abnormal contents.

Ultrasonographic Interpretation

On ultrasonography, the mucosa of the intestine appeared irregular. In Intestine there was appearance of hyperechoic sand deposits in the lumen of the intestine. The appearance of sand deposits and faecal contents was hypoechoic.

5. Foreign body

History: Dog was admitted with the history of engulfment of rubber ball during playing. The dog could not pass faeces afterward.

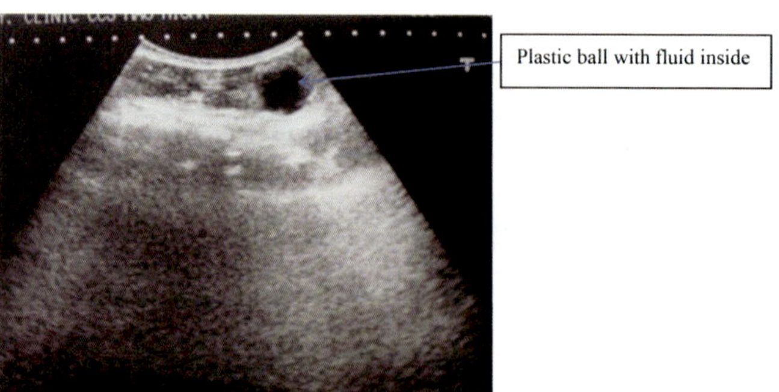

Fig. 126: is showing image of a plastic ball inside GIT of a young dog.

Ultrasonographic Interpretation

Ther e was appearance of anechoic rubber ball with hypoechoic margin of it. The anechoic image of the ball may be due to diffusion of fluid in the cavity of the rubber ball from the intestine. The rubber ball was surrounded by hypoechoic faecal material. The intestinal wall was hyperechoic may be due to inflammatory response.

Surgical Intervention

By enterotomy, the rubber ball was removed and the dog started passing faeces afterward. The recovery of the dog after surgery was uneventful.

ii) Foreign Body

History: Dog was presented with the history of engulfment of mango kernel two days back. Afterward the dog could not pass faeces .and there was constant vomition also.

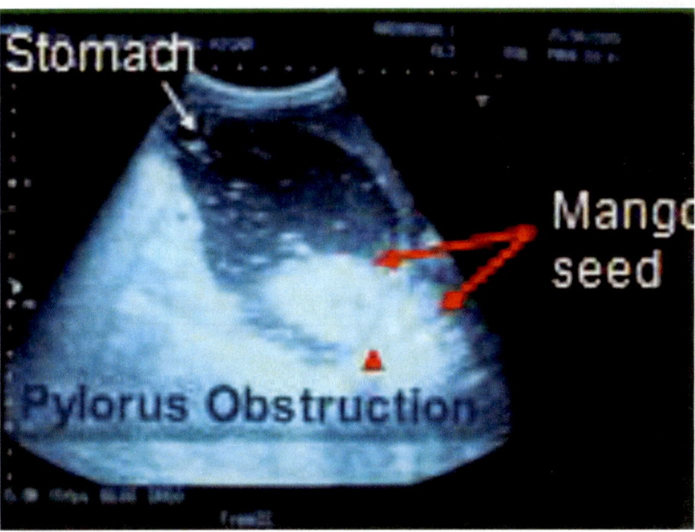

Fig. 127: is showing obstruction of pylorus with mango kernel in a dog.

Ultrasonographic Intervention

On ultrasonography of the stomach, the mango kernel can be seen as diffuse hyperechoic structure obstructing the pyloric end of the stomach partly present in the stomach and partly in the duodenum. The stomach can be seen as dilated structure full of anechoic fluid on the top and echogenic contents at the bottom.

Due to dense mango kernel in the pylorus and presence of fluid in the stomach, the acoustic enhancement can be seen.

Surgical Intervention

The mango kernel from the pylorus was removed by gastrotomy and further exploration of the intestine showed that another piece of the broken kernel was also present in the intestine so that half was also removed by enterotomy .This portion of the mango kernel could not be imaged by ultrasonography.

6. Atony of Intestine

History: Dog was admitted with the history of not passing faeces since fourteen days.

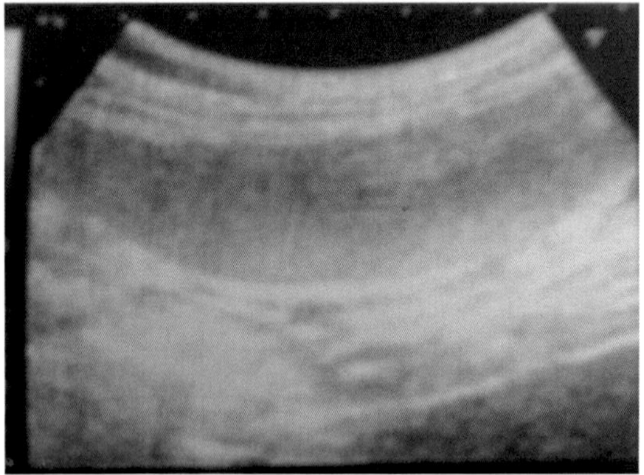

Fig. 128: is showing dilation of intestine in a dog and thining of intestinal wall of a case of not passing faeces.

Ultrasonographic Interpretation

The ultrasonography was carried out all over the intestine small as well large. There was dilatation of the intestine. The lumen of small as well as large intestine was full of hypoechoic faecal material.

Surgical Intervention

The enterotomy was carried out at two places First incision was given at the loop of small intestine (fig.) and second was given over the large intestine (fig.) The faecal material was evacuated through enterotomy incisions(fig.) and 50ml of paraffin was poured in the intestine. The intestinal wall and abdomen was sutured back .The dog passed the faeces after surgery and made an eventful recovery.

7. Intussusception

History: Dog was admitted with the history of not passing faeces since seven days. There was pain in the abdomen and crouching of the abdomen.

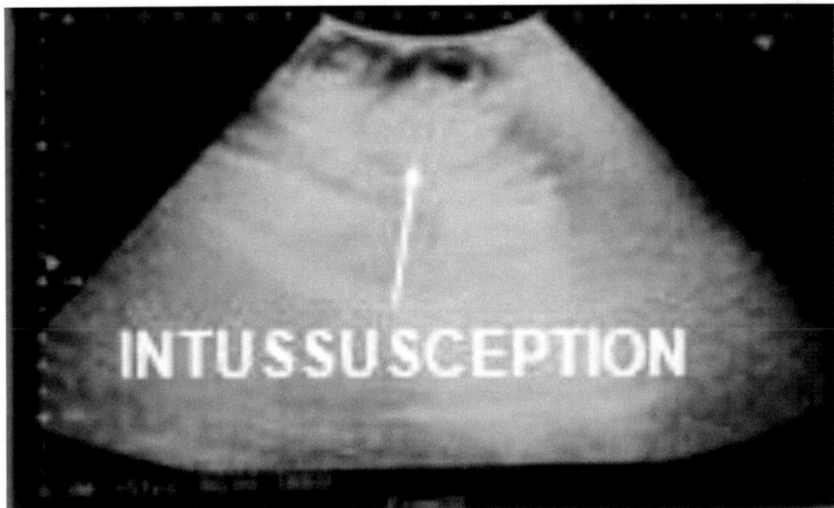

Figure-129: is showing image of intestinal intussuception in a dog.

Ultrasonographic Interpretation

Ther e was appearance of hyperechoic concentric layers of bowel wall within the hyperechoic intussuscepted segment. The hyperechoic appearance may be due inflammation. On the top of intussuscepted segment there was anechoic fluid filled portion (fig.). On another plane, the lesion was oval in shape. On both ends of the lesion, there was anechoic fluid filled portion and in the middle, the appearance was hypoechoic describing it as "target lesion".

Surgical Intervention

The intussuscepted portion of the intestine was resected out (fig.) and end to end anastomosis was done to maintain the patency of the intestine. The dog passed the faeces after surgery and responded well.

8. Retained Meconium in a Pup

History: -A pup was admitted with the history of not passing meconium after birth.

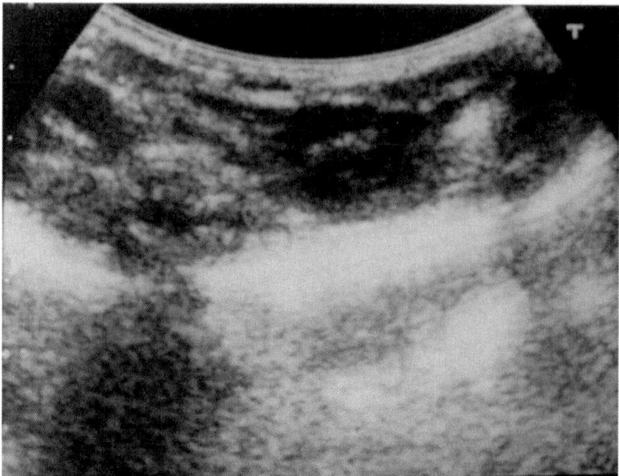

Fig. 130 is showing image of dilated intestine of young pup with history of retention of meconium

Ultrasonographic Interpretation

The ultrasonography of the intestine revealed that the small and large intestine was full of hypoechoic faecal contents and anechoic fluid contents. There was dilatation of the intestine. The wall of intestine was hyperechoic.

26

Case Reports

Case Report-1

A buffalo was having long standing problem of repeat breeding. On rectal palpation nodules were palpated along the uterine body. Following are the ultrasonographic images:

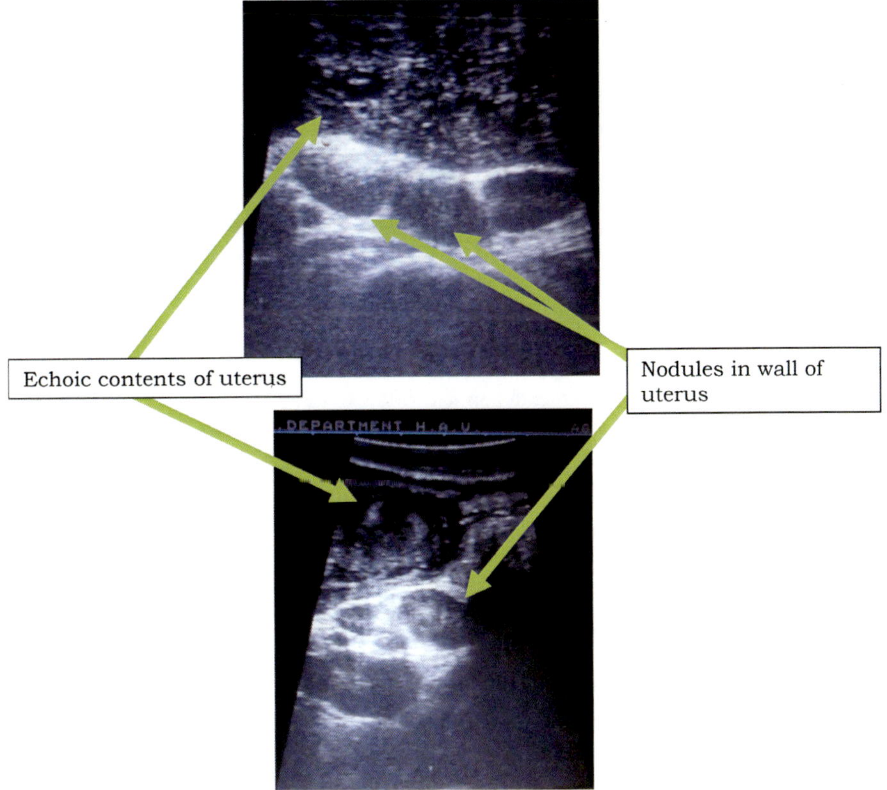

Fig. 131 & 132: are showing image of uterus with nodules and abnormal uterine contents.

Case Report-2

A buffalo was having long standing problem of anoestrus. On rectal palpation, the ovary was enlarged. On ultrasonographic examination the image of ovary was as below

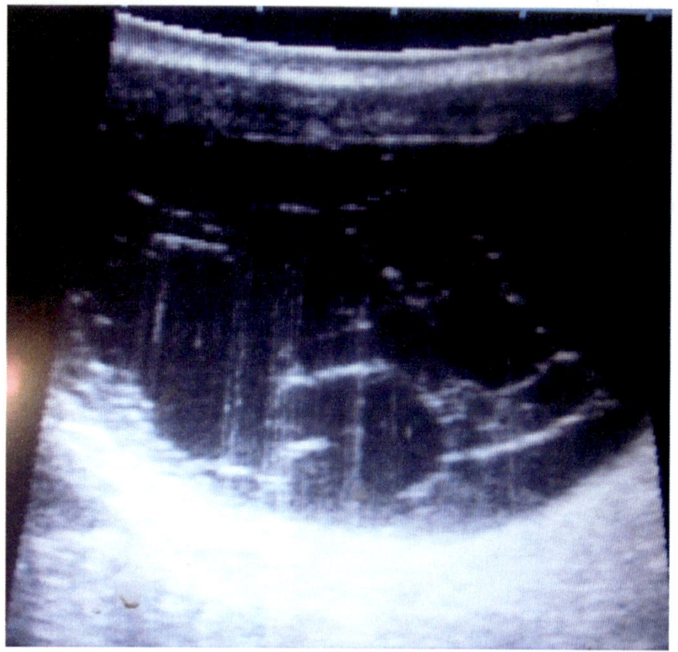

Fig. 133: is showing image of ovarian tumor in a buffalo. Note the compartmentalization of the image.

Case Report-3

(By Dr. Gyan Singh and R.K. Chandolia, 2015)

A veterinarian approached for ultrasonography of a bull with the history of blood in the urine. The veterinarian has tried various antibiotics, antihaemorrhagic, antiinflammatory drugs without any use. On ultrasonography the bladder of bull had damage to urinary wall with echoic casts in the urinary bladder.

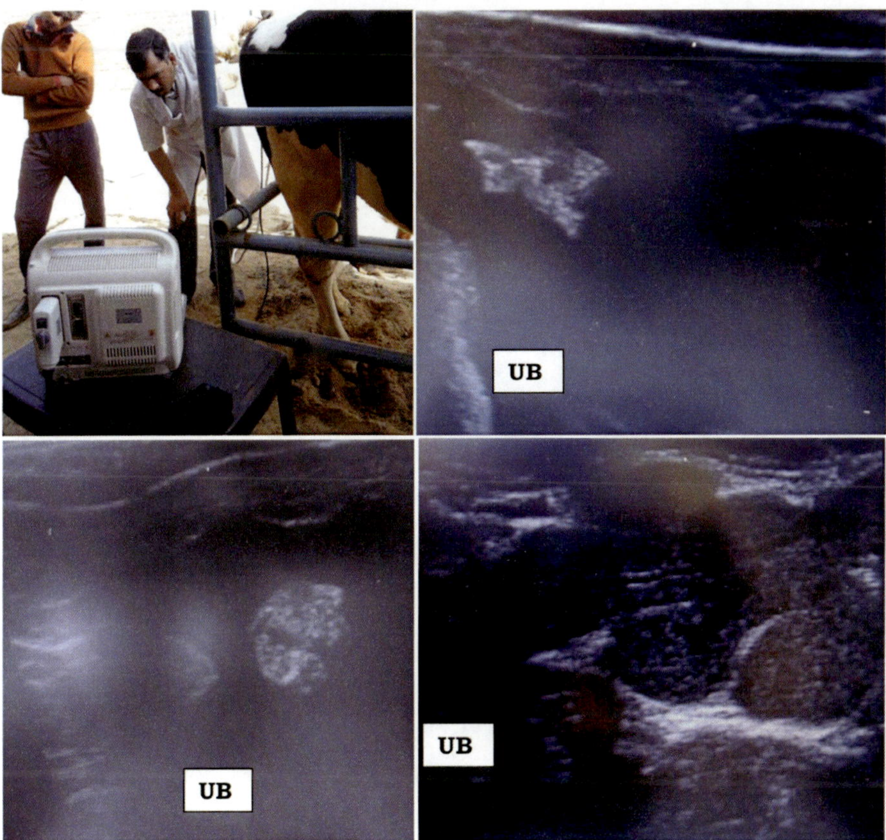

Photograph-34: and Figures 134-136 are showing scanning (top left) and echoic material in urinary bladder in three USG images in the above bull having blood in urine. The animal was fed mustard cake and the problem was solved after stopping feeding of mustard cake.

Case Report-4: Abnormal contents in uterus of a deer by Prem Singh and R.K. Chandolia 2015

A female Deer was presented with the history of bleeding from vagina. She was not allowing the male to mate and run away. The wild life personnel wished to know, what was the cause of bleeding?

The ultrasonography revealed that there was nodular contents in the uterus as shown in picture. Various segments of uterus were seen. The images as shown below were taken from the vicinity of urinary bladder.

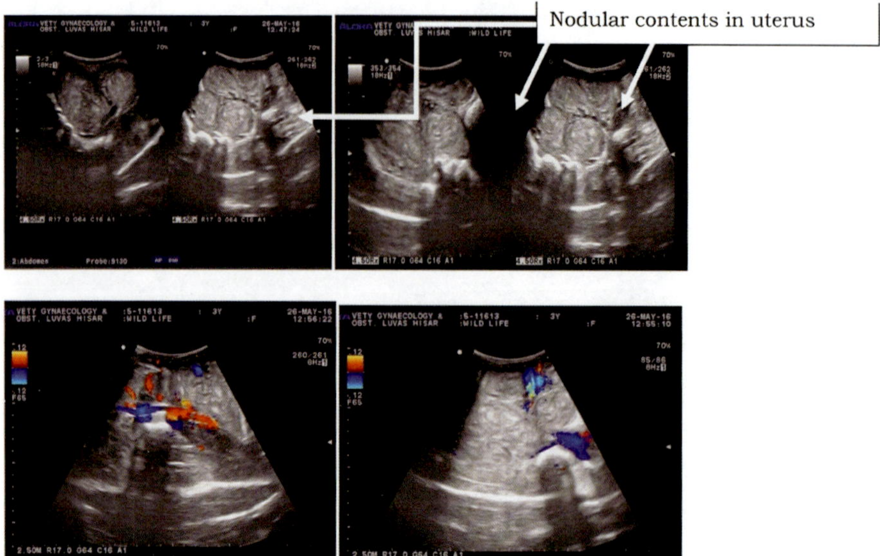

Fig. 137 & 138: is showing cross sectional images of area in the vicinity of urinary bladder. There was nodular appearance (above panel) and increase in blood supply (lower panel) of the area in the female Deer.

Case Report- 5: Testicular tuberculosis in a bull

(Prem Singh and R.K. Chandolia)

A adult cow bull was presented with hardness and enlargement of one testis. On palpation the enlarged testis was hard like stone. Other tesis was also hard but not like other testis. On ultrasonography the typical lesions of tuberculosis were seen. The case was confirmed by pathological examination as case of tuberculosis.

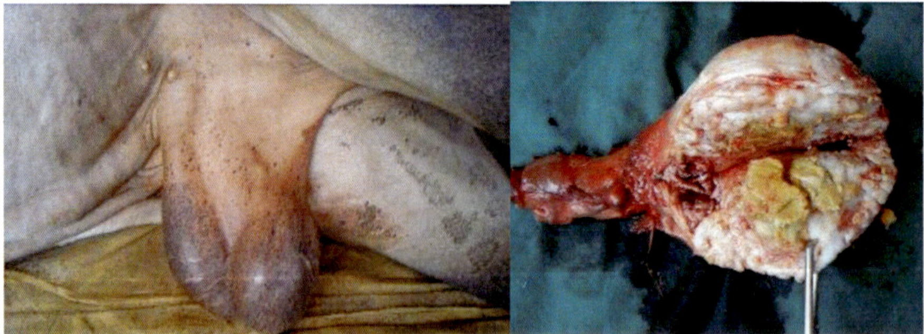

Photograph-35 & 36 are showing a bull suffering from testicular tuberculosis. One tesis was enlarged (left panel) that was very hard. Affected testis (cut open) on right side.

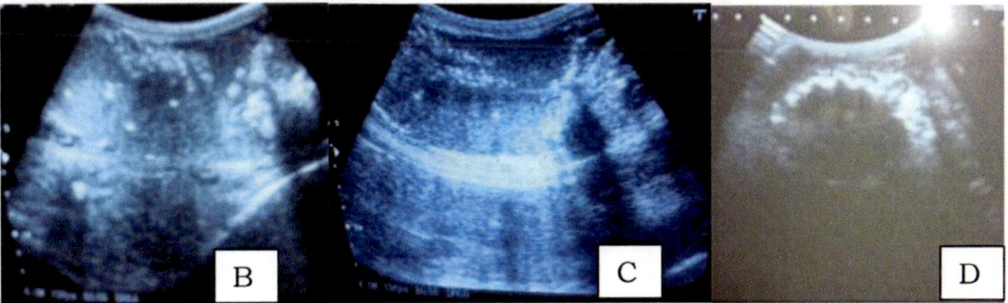

Figure-139 is showing USG image of affected enlarged testis (B) and companion tesis (C). Wall of companion testis also showed echoic lesions (D).

Case Report 6: Testicular Inflammation in a Buffalo Bull

A buffalo bull of Murrah breed with good physical conditions was presented for ultrasonography. The bull was prize winning male and owner reported that the semen of this buffalo bull was in great demand. The age of the bull was 6 years. The owner reported that the bull had swelling of the testicles. On examination it was found that one testis was smaller while other one was enlarged and hot to palpate. It was thought a case of Unilateral-Orchitis. On palpation the testis was hard. Ultasonography was done using Intraoperative probe (Toshiba, India) and Toshiba nemio-XG ultrasound machine.

Observations: One testis was hyperechoic, but size not enlarged. The other testis was enlarged with pockets of fluid. There were compartments in one side i.e. anterior longitudinal side of testis. See the pictures below:

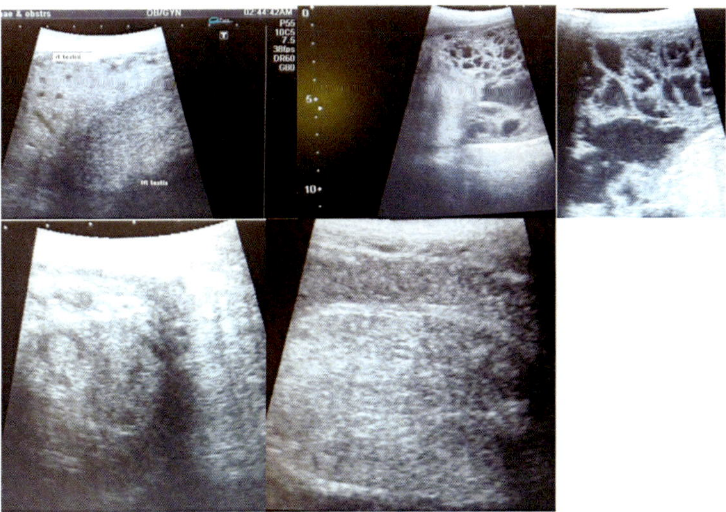

Fig. 140 &141: is showing abnormal images of testis from different angles. There is fluid in upper portion of testis (upper panel) and image of lower portion of testis in buffalo bull.

Case Report -7: Fetal Mortality in a Buffalo

Three buffaloes of a farm were presented for pregnancy diagnosis. The animals were mated almost two months back. One buffalo has fetal death see picture at sr. No. (A) and others had normal developing fetus shown in picture at no. (B).

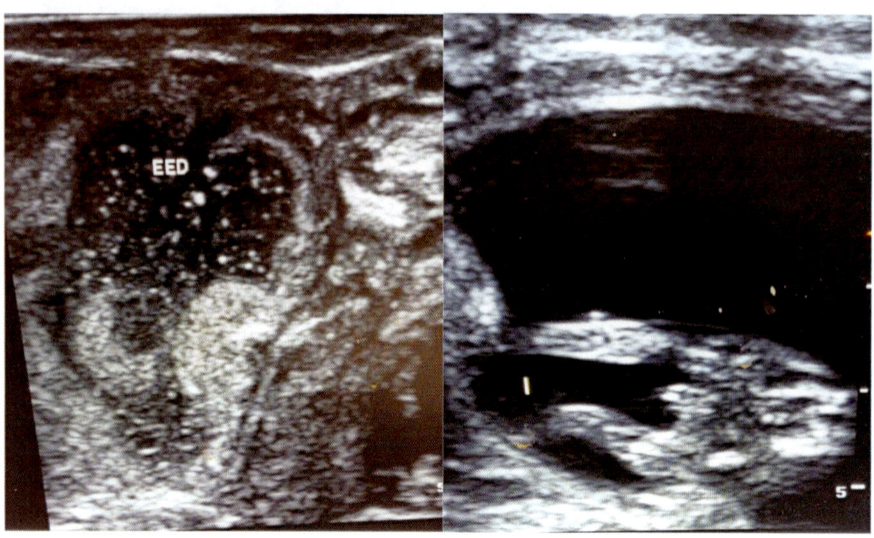

Fig (142) **Fig (143)**

Fig. 142 & 143: Figure on left (140) is showing fetal mortality in which fetal fluid is cloudy and abnormal shape of fetus, while figure 141 is showing normal fetus in which fetal fluid is clearly anechoic showing all parts of fetus. In normal fetus heartbeat was also seen. From echoicity of fluid and fetal movement fetal mortality can be easily diagnosed.

Case Report -8.

A bitch (named Juli) of 7 years age was mating 53 days back. The owner reported foul smelling discharge. The bitch was dull. On ultrasonography, urinary bladder was empty. The uterine horns showed compartments. One compartment contained fluid filled with abnormal echoic fluid. Other compartment showed cross wheel type with some pockets of unechoic nature. It is suspected that the bitch got implantation and at certain stage, uterus got infection and pups died, got absorped leaving behind these pockets. See images

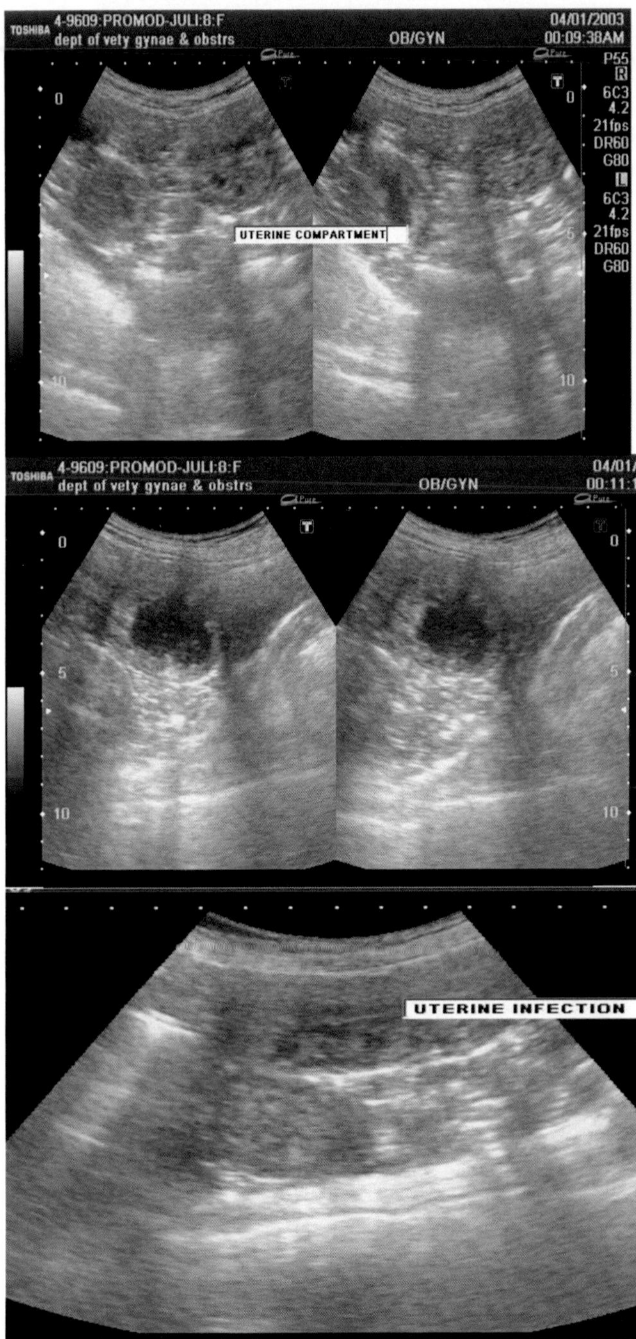

Fig. 144 & 145: are showing uterine compartments with thick wall having contents in some sections

Case Report-9 A case Report: Reticular abscess

A dog was presented with eye problem. Ultrasonography was conducted with available curvi-linear probe and following image was obtained and diagnosed as a case of reticular abscess. See thickness in retina. The case was confirmed with surgical procedure.

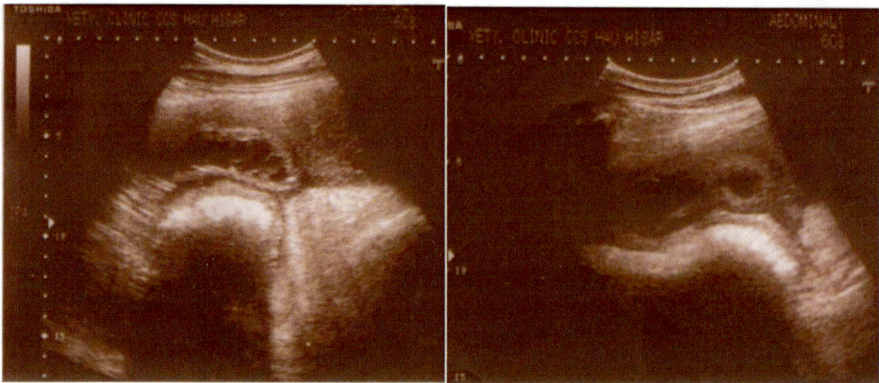

Fig. 146 & 147: are showing images of reticular abscess in a dog.

Case Report-10

A buffalo was not conceiving and showing sign of heat for 3 to 4 days. In absence of ultrasonography working veterinarian inseminated the animal at two occasions but animal failed to conceive. The animal was subjected to ultrasonography on next oestrus. Following was the USG image of one ovary.

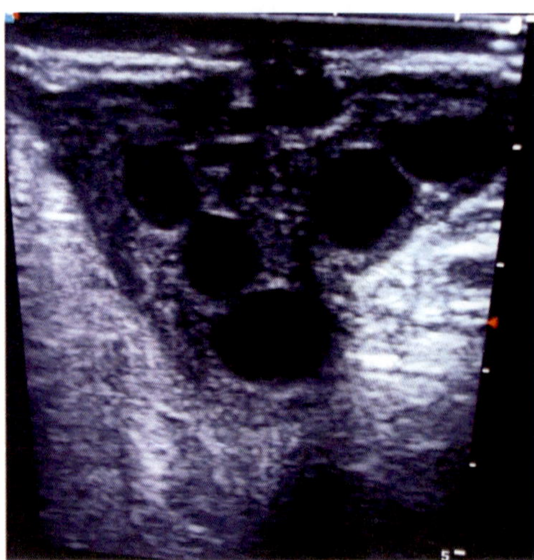

Fig. 148: is showing cystic ovary in a buffalo.

Case Reports 11 & 12

Two cases were presented with the history that pet was not taking food. In one case abdomen appeared normal, while in other one the abdomen was enlarged. On ultrasonography the cases were diagnosed of Ascitis. USG images of ascitis cases are shown below:

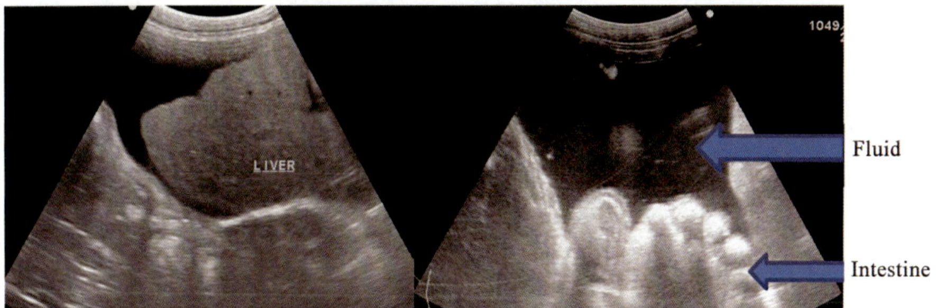

Case Report 13 & 14

These two case were presented for ultrasonography. History of one case indicated problem in urine output in the dog. The history of other case was blood in urine. Following are the ultrasonograms of these cases. In one case there was pus in bladder, while other one had large stone (cystolith) in the bladder.

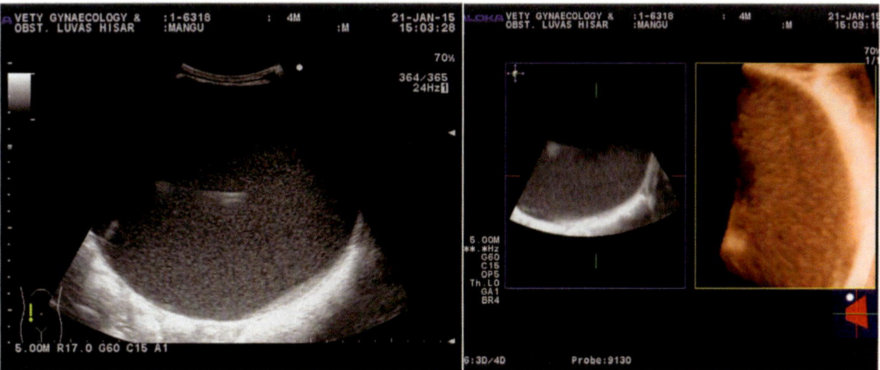

Fig. 150: is showing pus in the urinary bladder of a dog

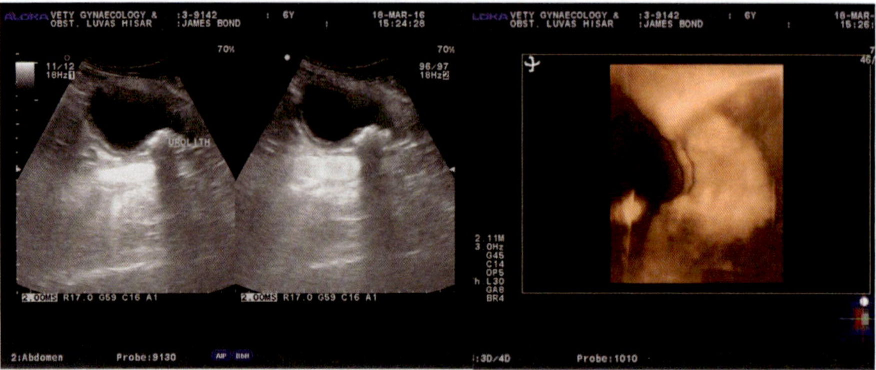

Fig. 151: are showing 2D & 3D view of stone in urinary bladder (Cystolith) in a dog.

Case Report 15: Microlithiasis in a Bull

A bull, age 6 years, had small testes and was examined ultrasonographically. Body score was poor. The testes showed hyperechoic microlithiasis i.e. reflective deposites.

History of the case: A buffalo bull was scanned and found to contain echogenic fine nodules in both the testis, age of bull about 6 years. These are considered as calcium deposites that disturb testicular function. In human this has been linked to testicular cancer and other testicular problems.

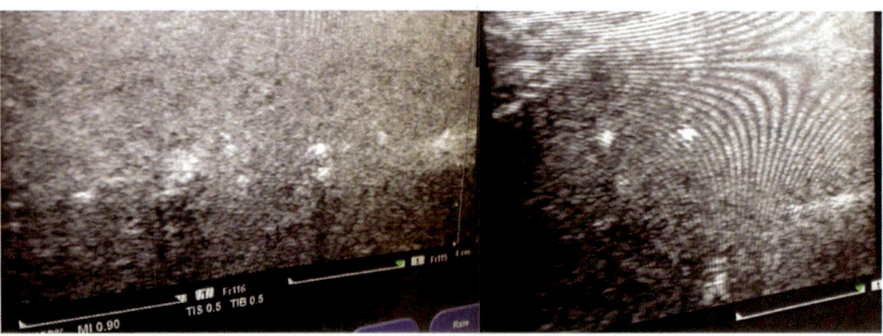

Fig. 152: is showing "Hyper-reflective (white) tubercles in testis of buffalo bull (Microlithiasis).

Case Report-16: Sperm Ball in Urine of Monkey

Two Monkeys were presented with the history of not taking food for last three days. There was no temperature. There were no other obvious signs of illness. The monkeys were subjected to ultrasonography. It was observed that urinary bladder was full of urine and there was some thing that blocked urine. The image was different than usual cystolith. It was decided to go for surgery to remove the obstruction. The monkeys were operated as per standard procedure. After opening the urinary bladder, the cause of obstruction was removed. It as roung structure, but was not too hard. It was sent to lab for analysis. The report presented was that it was a ball of sperms. As shown in the pictures below the sperms were separted and photographed.

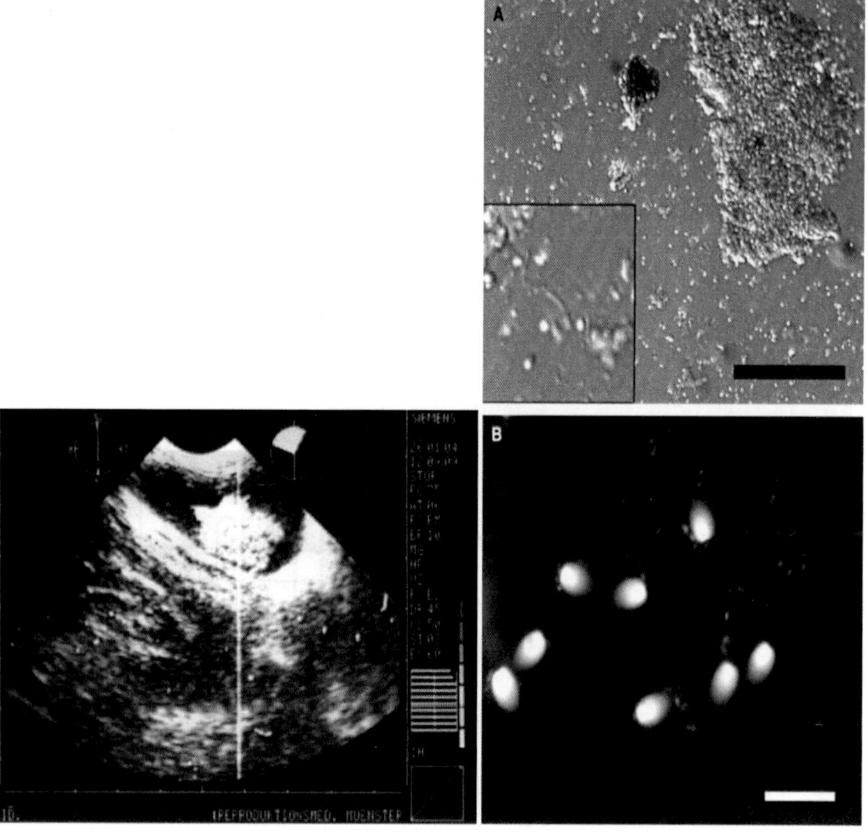

Fig. 153: is showing presence of blocking irregular mass in the urinary bladder **(left)**, analysis of ball and photo of sperm **(upper right)** and illuminated pictures of sperms **(lower right)** in the monkey.

Case-17: During routing ultrasound examinations for pregnancy diagnosis at an organized animals farm, one buffalo was diagnosed with Mucometra.

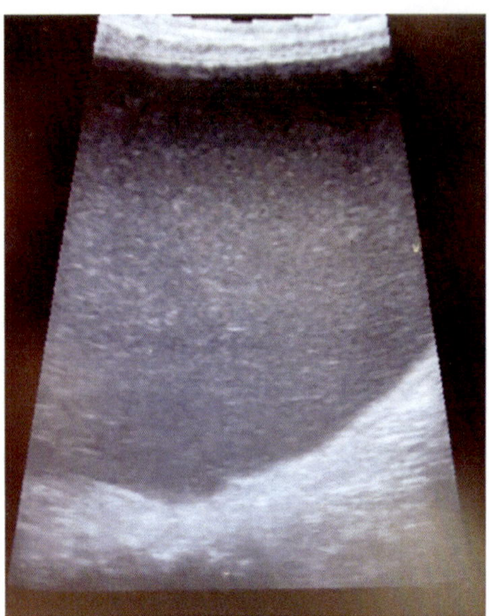

Fig. 154: is showing image of mucometra case in a buffalo

Case -18: Two buffaloes were diagnosed with pyometra during examination of animals on a farm. Followings were the ultrasonographic images.

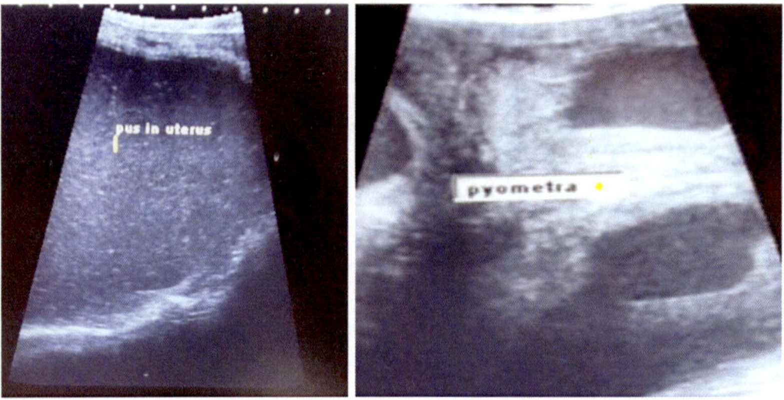

Fig. 155 & 156: are showing images of pyometra in buffaloes.

Case-19

A bufffalo calf was presented with the history of not passing uring since two days. The calf had urine blockade 20 days back. The local veterinarian performed tube cystotomy. Few days later, the calf dropped the catheter leading to non-passing of urine. On ultrasonography mess like structures were seen in the urinary bladder as shown below.

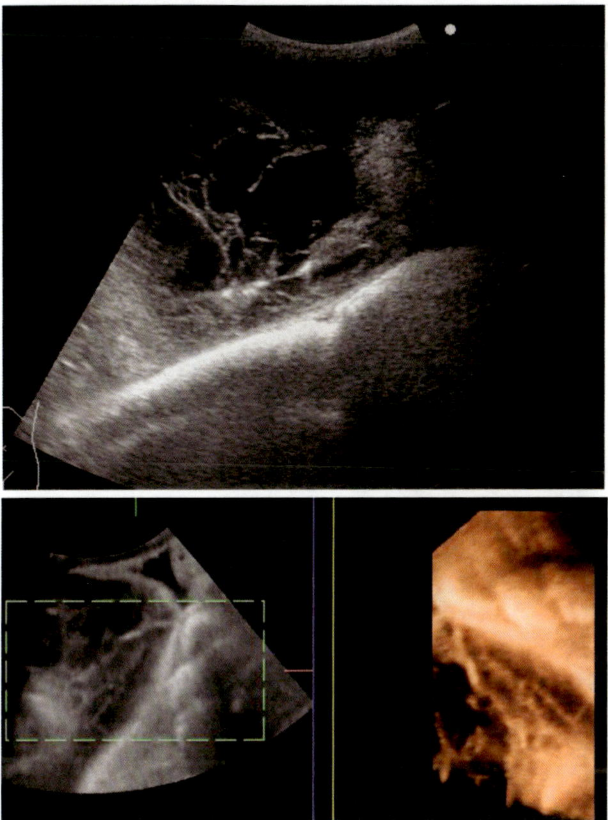

Fig. 157 & 158: are showing complecations of tube cystotomy in a buffalo calf.

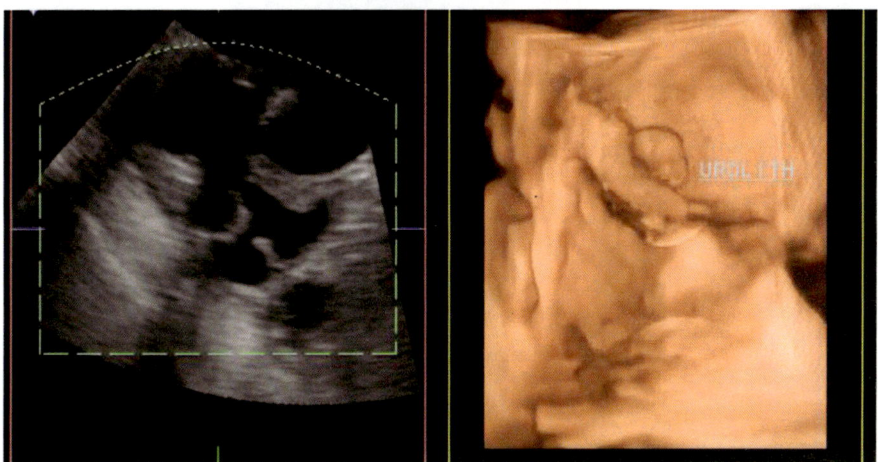

Fig. 159: is showing urolith clearly in 3 D image (right) in a dog

Case No.- 20

A newly born buffalo calf was presented with history of not passing urine. On ultrascan presence of patent urachus was diagnosed.

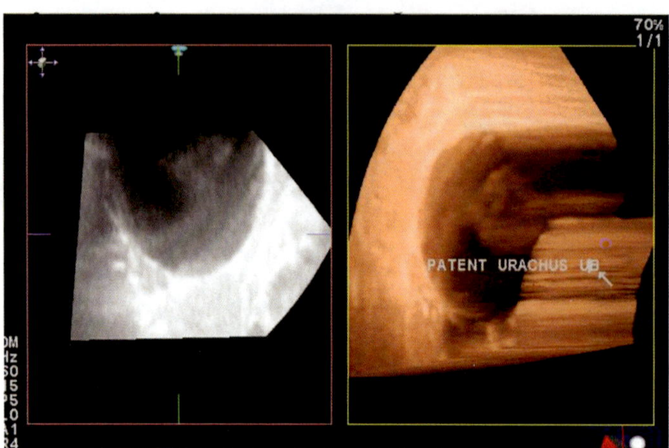

Fig. 160: is showing internal structure of urinary bladder of patent urachus. A covering on internal surface of urinary bladder was seen along sith a raised passage.

Case-21: A street dog was presented for ultrasonography as dog was weak and not eating properly. Following ultrasound images of abdominal organs was obtained. The liver has a thick area. The spleen was enlarged. The kidney also showed abnormal structures indicating nephritis.

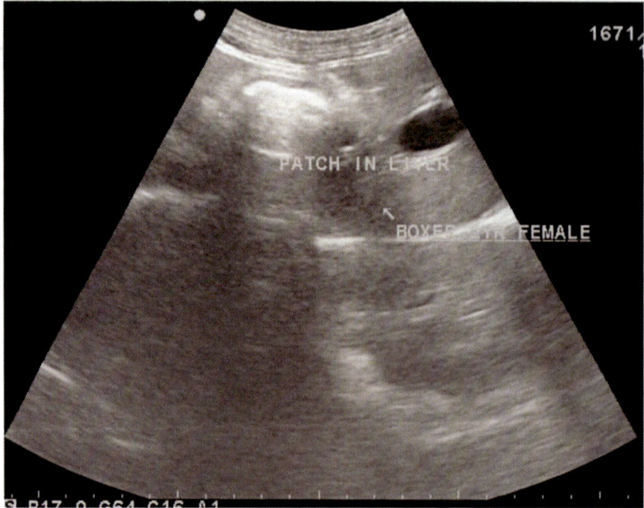

Fig. 161: is showing an ecnoic patch in the liver near gall bladder in street dog

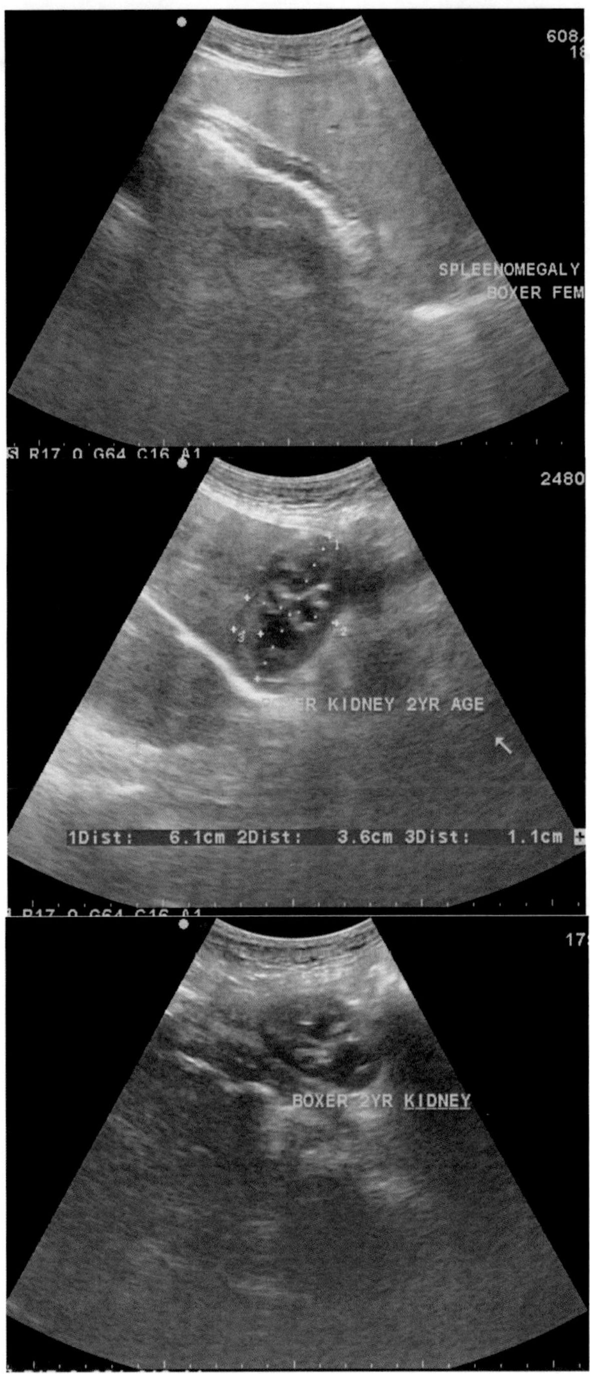

Fig. 162-164: are showing enlarged spleen **(top)**, left kidney **(middle)** and right kidney **(bottom)** in a street dog.

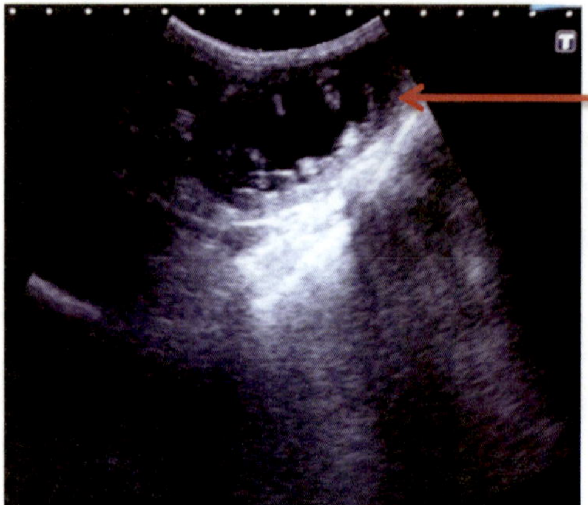

Case 22: Ultrasound image of Hydronephrosis in a dog. Arrow is showing fluid accumulation the kidney.

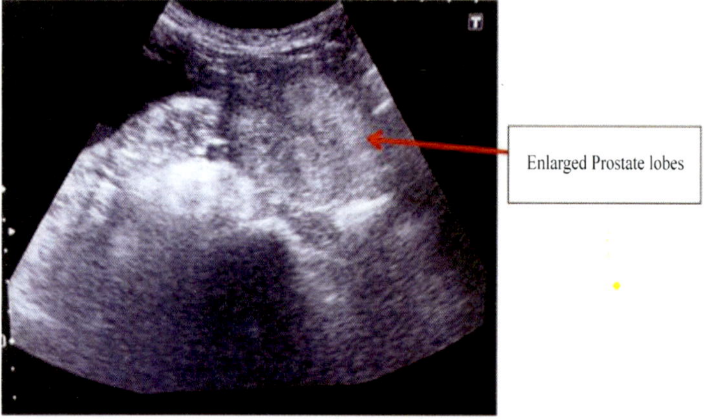

Case 23: A dog was having difficulty in urination. On ultrasonography enlarged prostate was diagnosed.

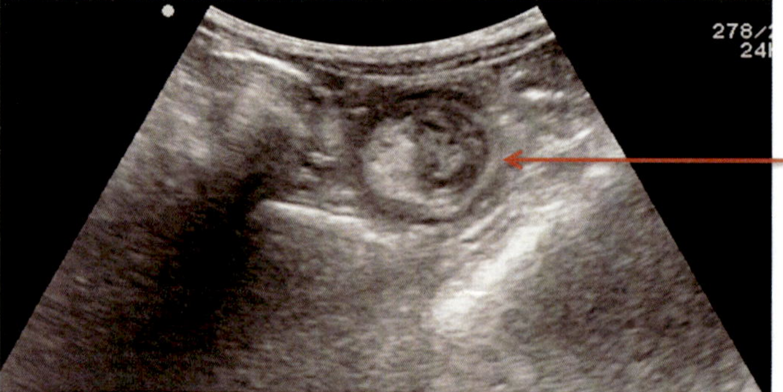

Case 24: A marriage ring of a couple was lost. The ultrasonography showed image of ring (arrow) in the intestine of their dog.

27

3D-Ultrasound Images Taken Sequencially for Fetal Development in Dog

3D-Ultrasonographic Images from Bitches: Fetus development on different days of gestation (Jinsa George and R.K. Chandolia, 2012)

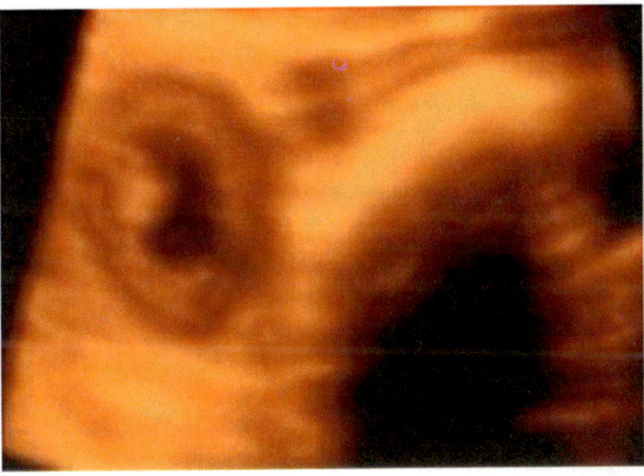

Fig. 165: is showing 3D-image of conceptus on Day 20 of pregnancy

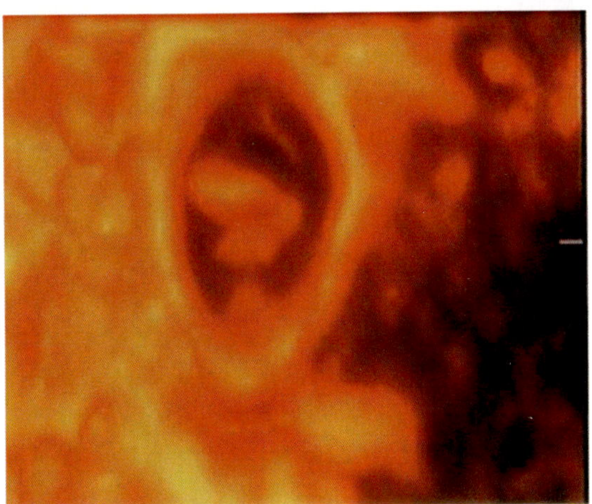

Fig. 166: is showing 3D-image of conceptus on Day 21 of pregnancy

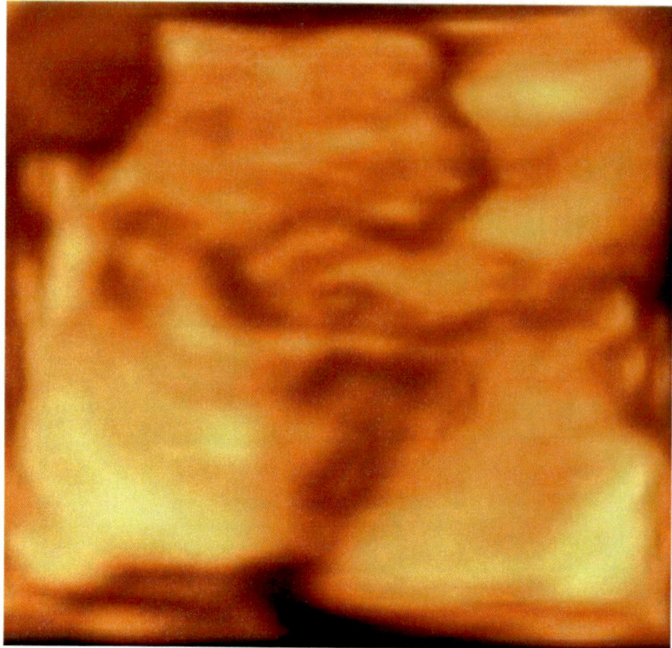

Fig. 167: is showing 3D-image of pup on Day 30 of gestation

28

Ultrasonographic Images of Goat Fetus in Utero and Its Internal Orgnas from Day 24 Onwards

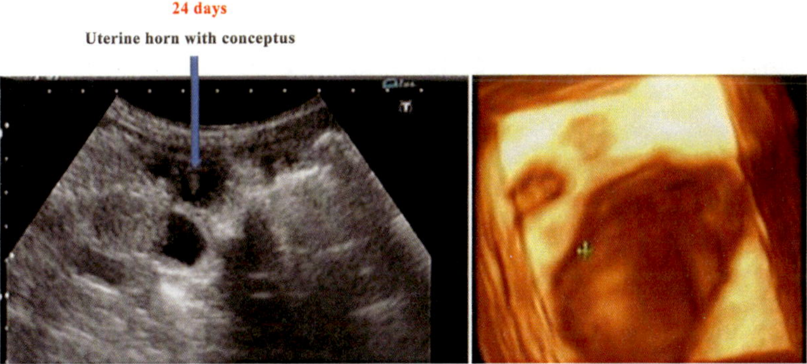

Fig. 168 & 169: are showing goat-conceptus on day 24 of gestation in 2D **(left)** and 3D ultrasonography **(right)**.

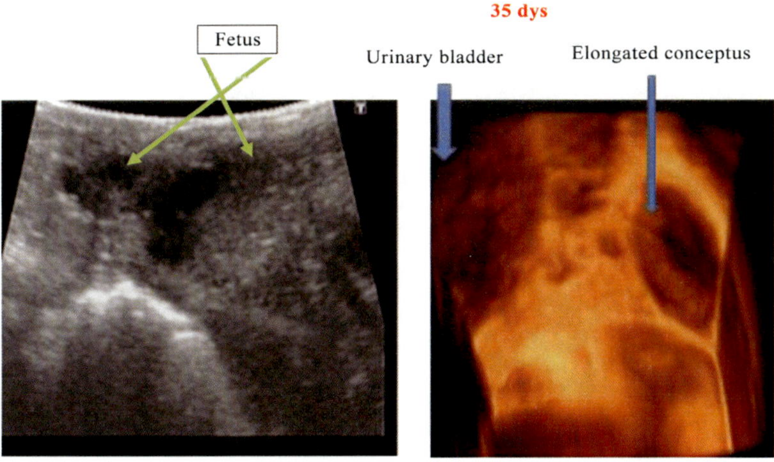

Fig. 170 & 171: are showing goat-conceptus on day 35 of gestation in 2D **(left)** and 3D ultrasonography

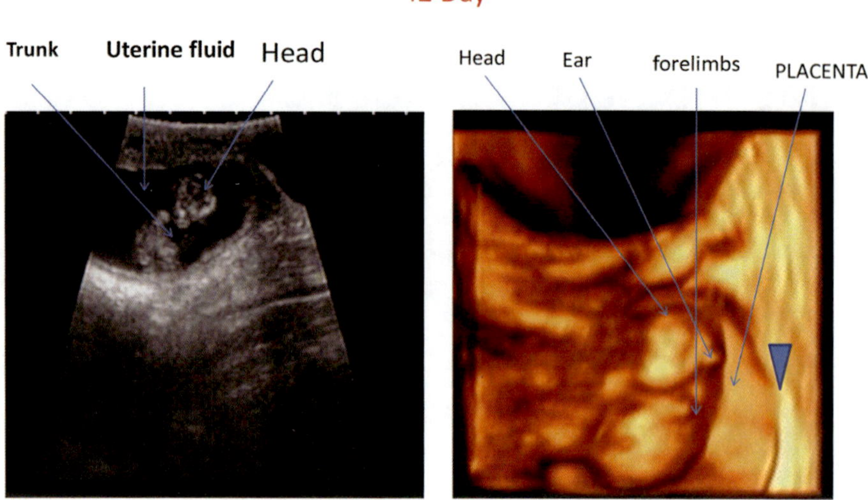

Fig. 172 & 173: are showing goat-conceptus on day 42 of gestation in 2D **(left)** and 3D ultrasonography

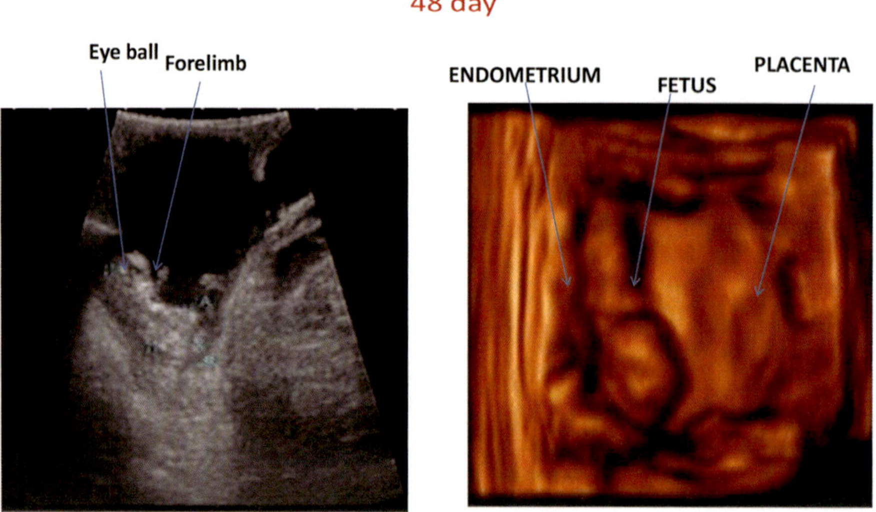

Fig. 174 & 175: are showing goat-conceptus on day 48 of gestation in 2D **(left)** and 3D ultrasonography

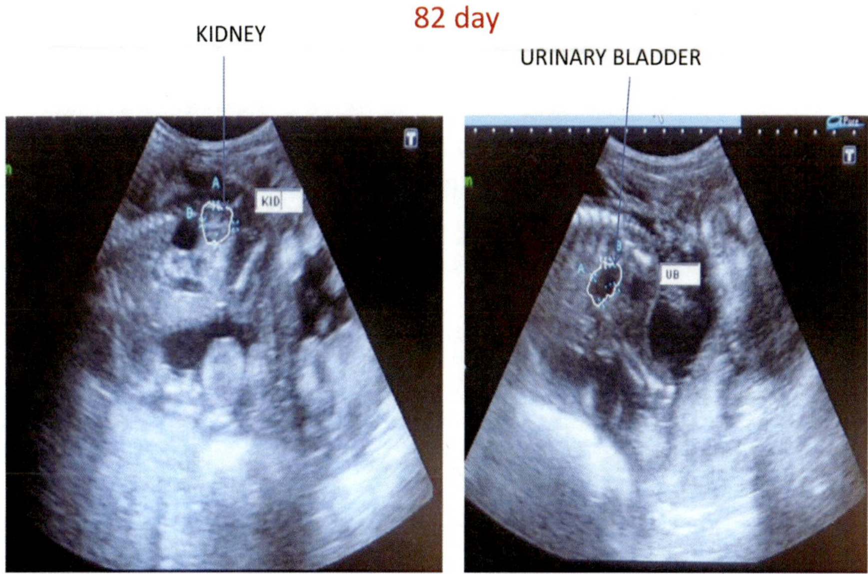

Fig. 176 & 177: are showing goat-conceptus on day 82 of gestation in 2D **(left)** and 3D ultrasonography

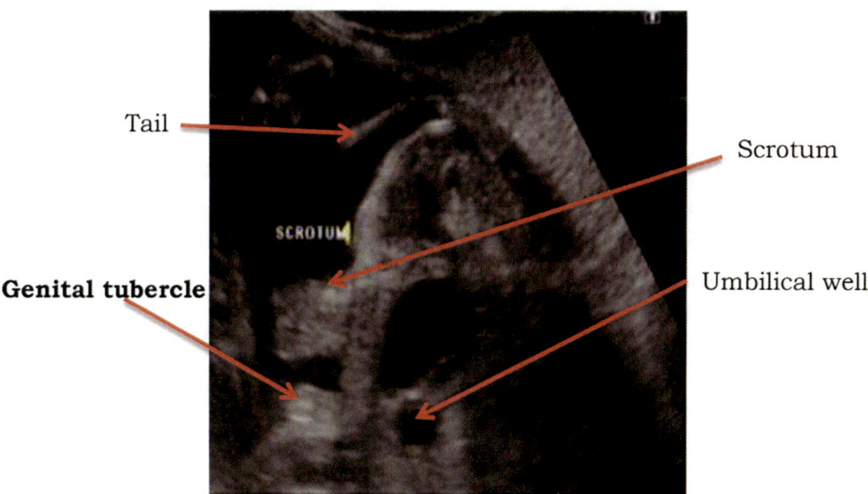

Fig. 178: is showing organs of goat fetus on day 90

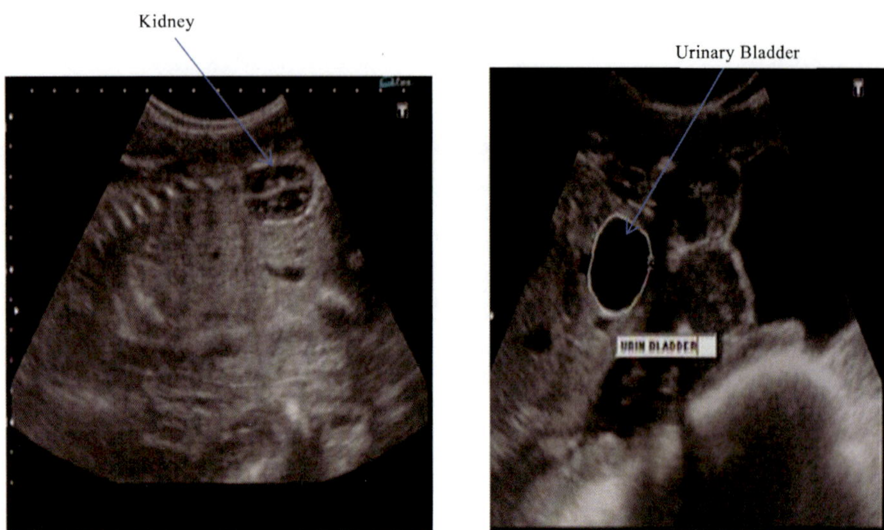

Fig. 179: is showing internal organs of goat-conceptus in utero at 3 month of gestation.

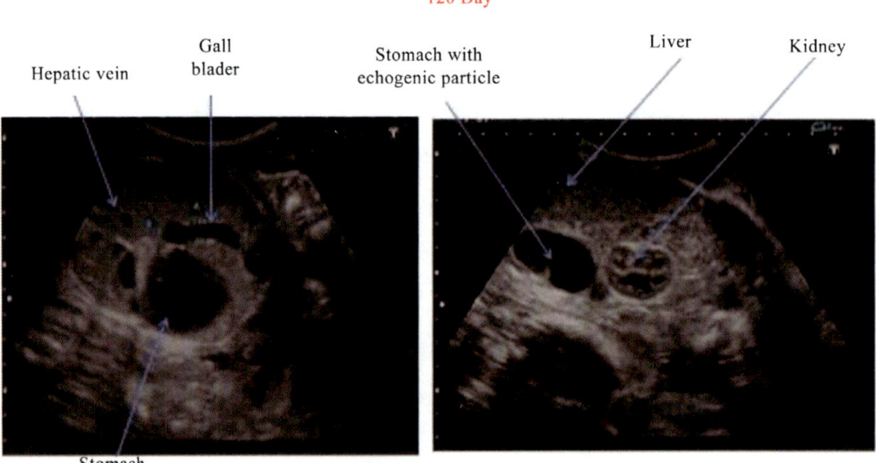

Fig. 180: is showing internal organs of goat-conceptus in utero at 4 month of gestation.

Blood flow in umbilicus on different days of pregnancy in goat

39 Day

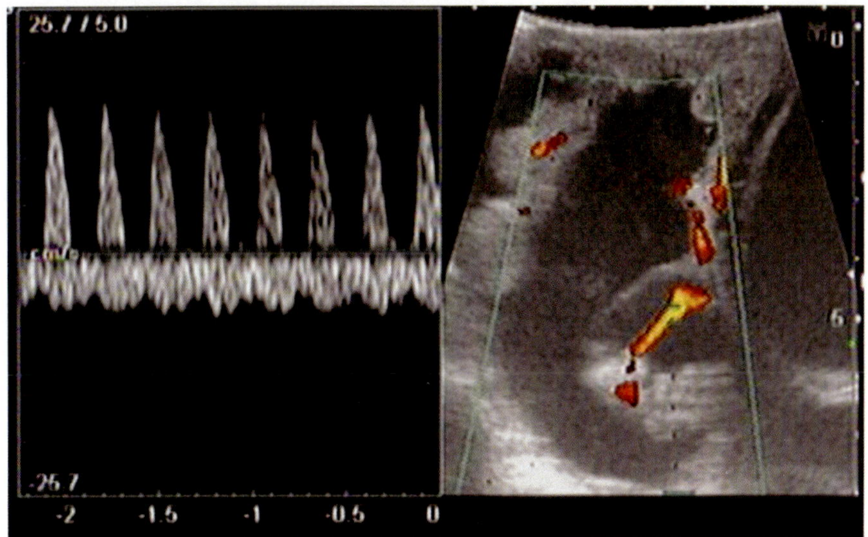

Fig. 181: is showing blood supply to conceptus of a goat on day 39. The tooth like upward projections are depicting arterial supply while downside is venous pattern.

48 day

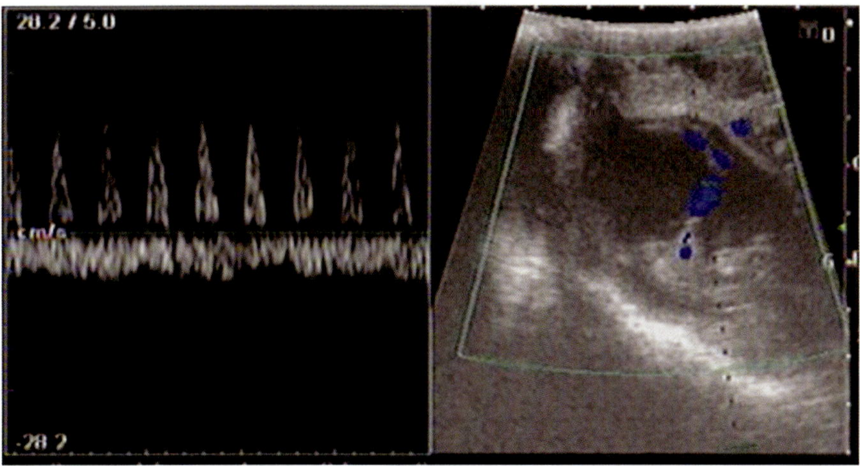

Fig. 182: is showing blood supply to conceptus of a goat on day 48. The tooth like upward projections are depicting arterial supply while downside is venous pattern.

67 Day

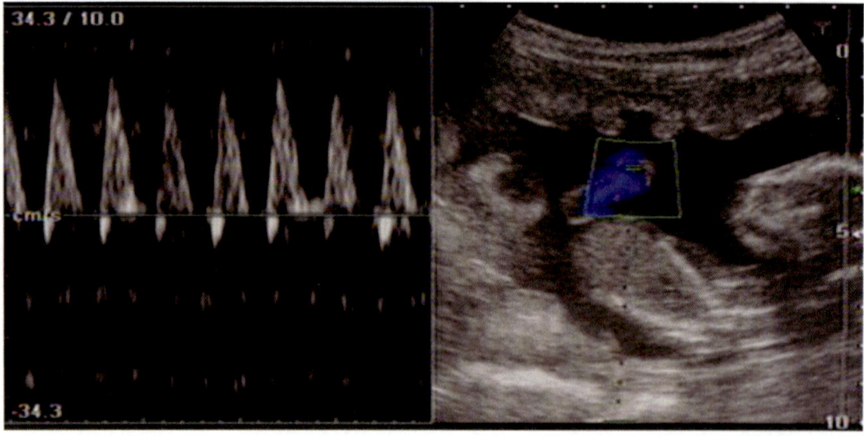

Fig. 183: is showing blood supply to conceptus of a goat on day 67. The aliasing effect is seen where tip of peak are cut and are shown below base line. Similarly split image of venous supply is also seen.

82 Day

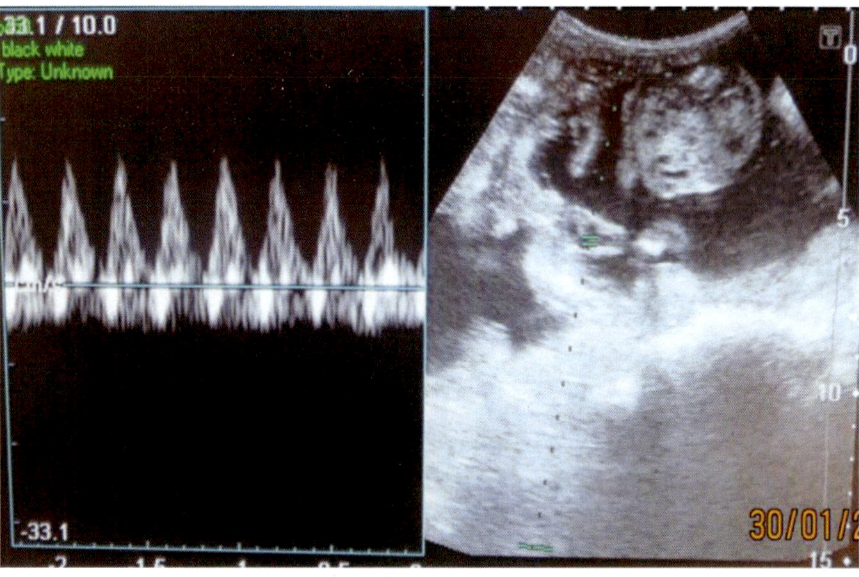

Fig. 184: is showing blood supply to conceptus of a goat on day 82. The gap between arterial peak is filled with diastolic small peak in between long systolic peaks.

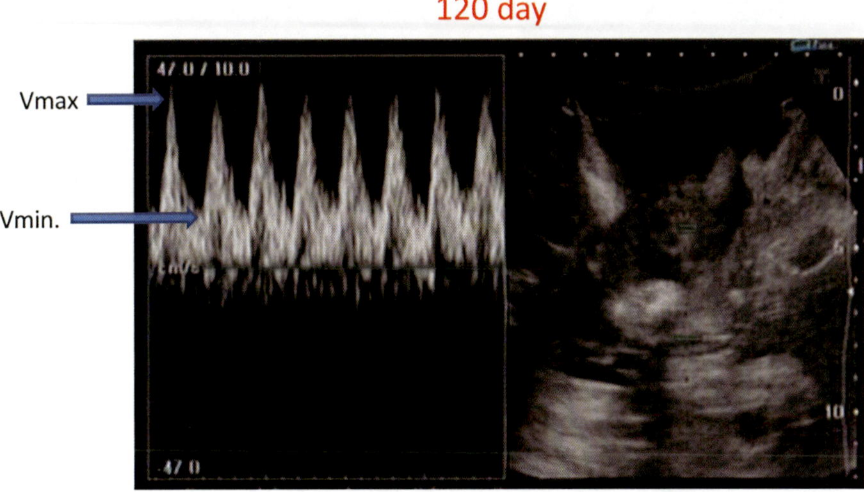

Fig. 185: is showing blood supply to conceptus of a goat on day 120. Both systolic and diastolic peaks of umbilical aretery are seen above base line and pattern of umbilical vein downside.

29

Scanning Positions and Different Views of Organs in Dog from Different Angles

1. Scanning position of urinary bladder: Urinary bladder is imaged in dorsal recumbency keeping in view the topographic position of the urinary bladder just cranial to the pubis. Position of good image of urinary bladder also depends upon amount of fluid in bladder. Full urinary bladder can be easily viewed; however there is difficulty in getting image of empty bladder.

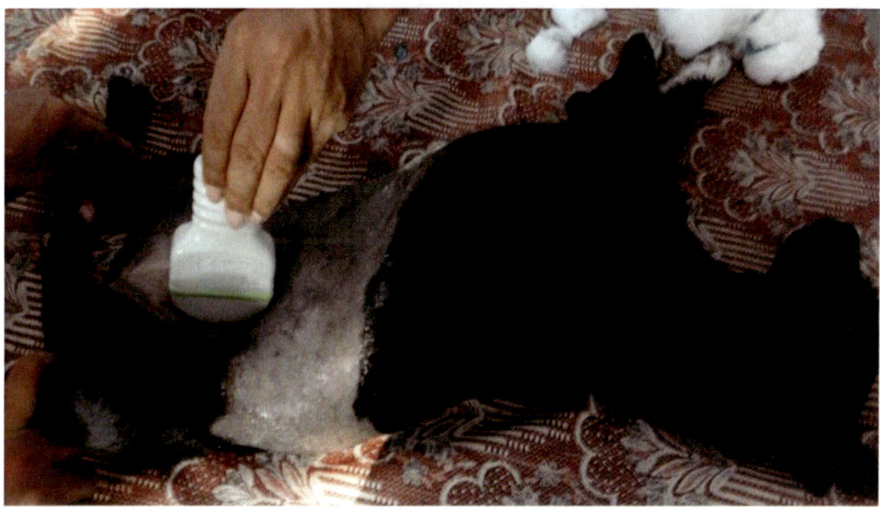

Photograph-37: is showing position of probe for scanning of urinary bladder in a dog

Images of Urinary Bladder in Different Pathological Conditions

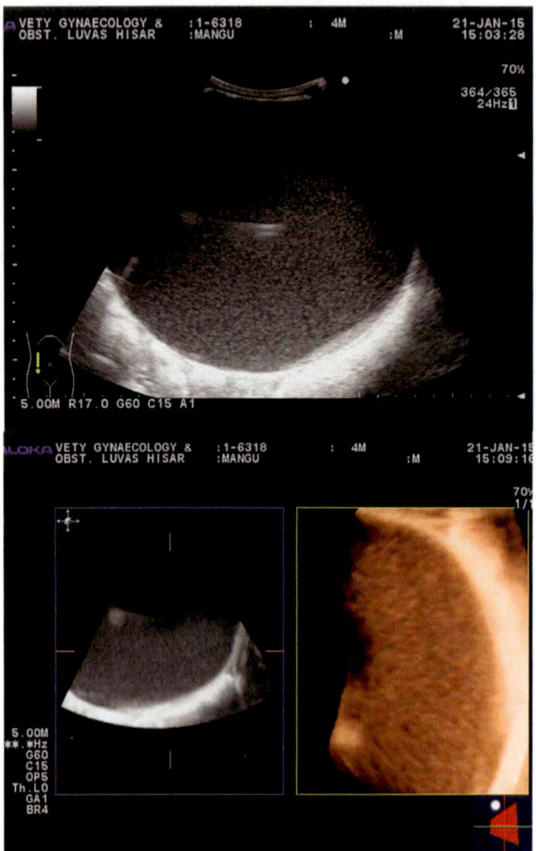

Fig. 186: is showing pus in the urinary bladder with 2D and 3D scanning

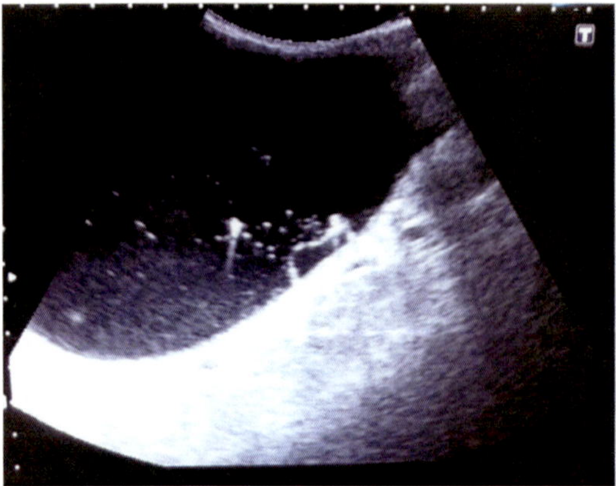

Fig. 187: is showing cystoliths with pus in urinary bladder in a dog

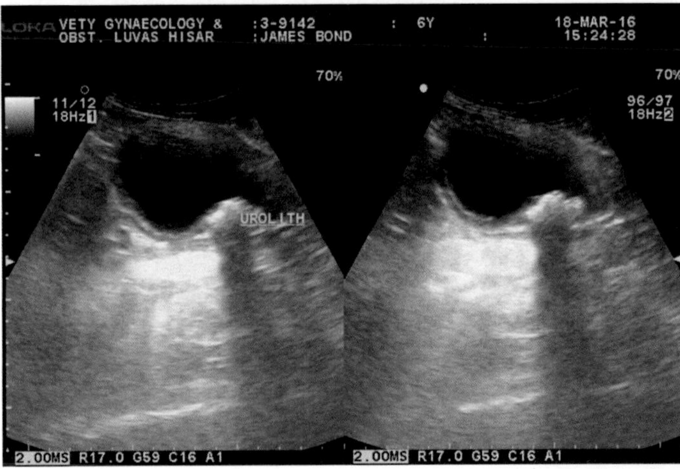

Fig. 188: is showing stone in urinary bladder (uroliths) and shadowing under urolith in a dog.

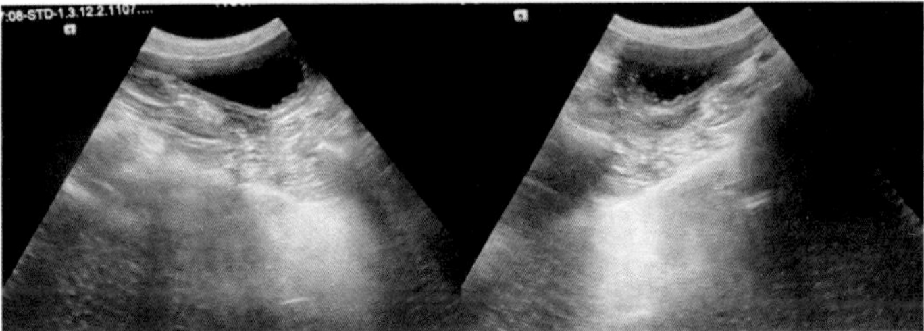

Fig. 189: is showing cystitis (lesions in the wall of UB) in a dog

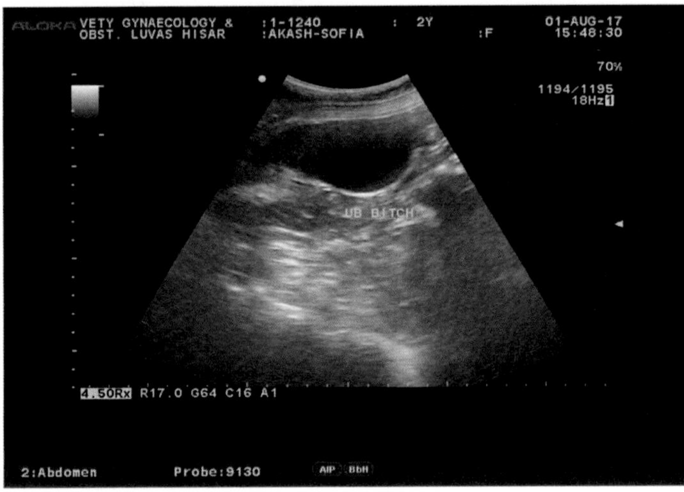

Fig. 190: is Showing normal urinary bladder

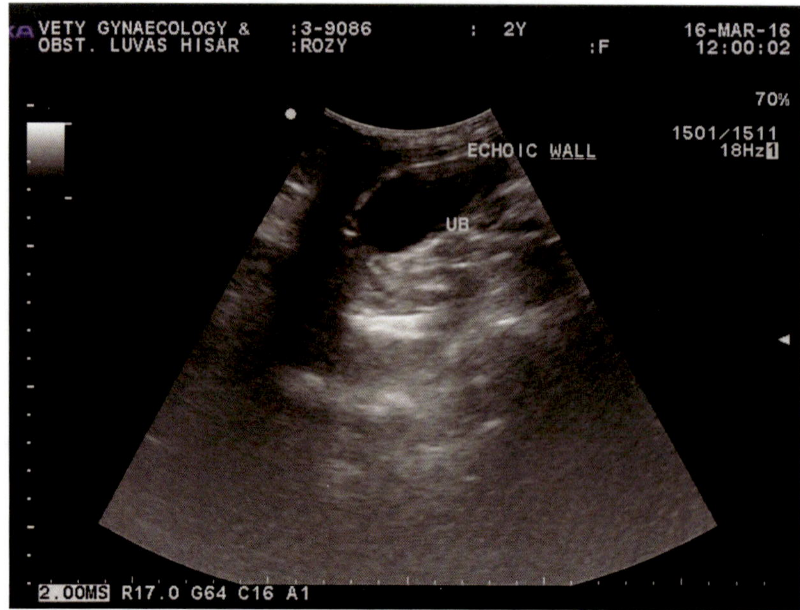

Fig. 191: is showing urinary bladder with echoic wall in a dog.

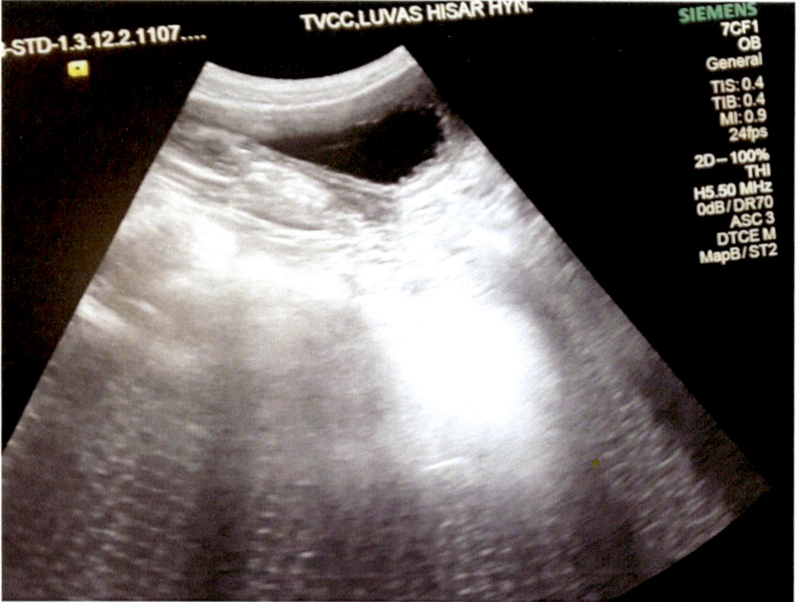

Fig. 192: is showing abnormality of urinary bladder; wavy margin of the wall of urinary bladder is seen.

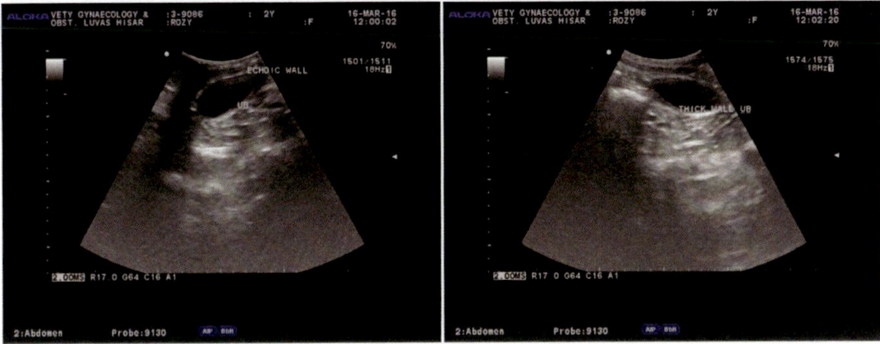

Fig. 193: is showing echoic wall of urinary bladder

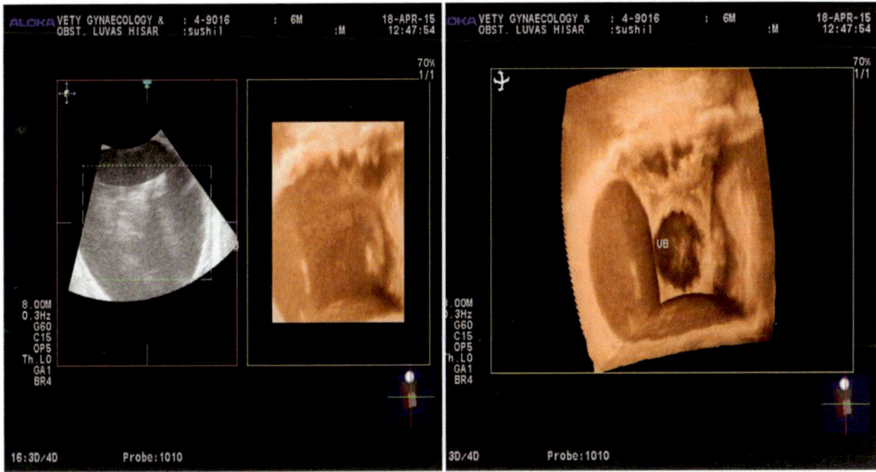

Fig. 194: is showing abnormalities of urinary bladder

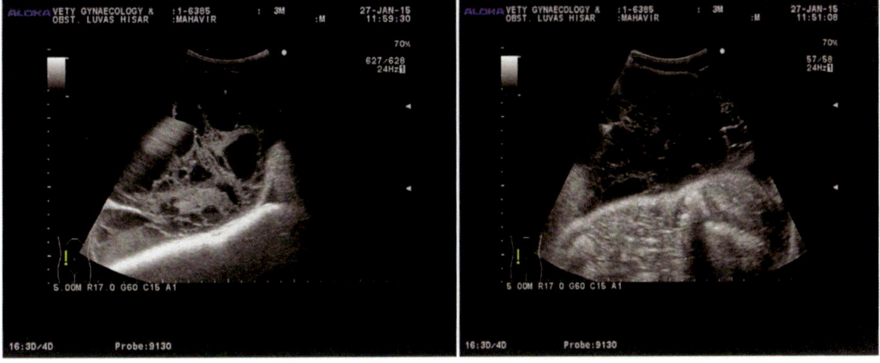

Fig. 195: is showing side effect of catheterization of urinary bladder in a case of tube-cystotomy.

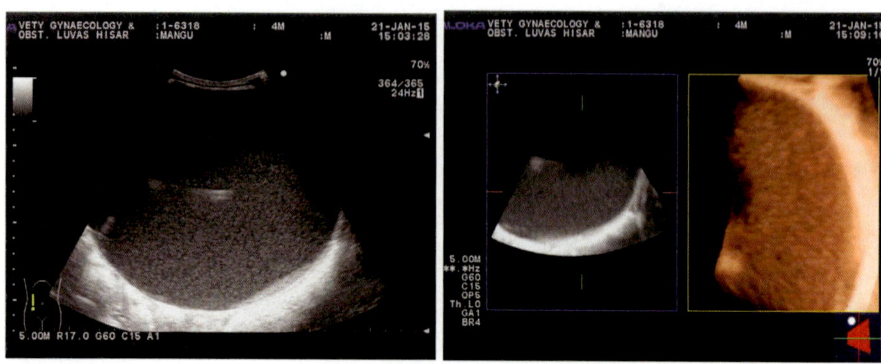

Fig. 196: is showing 2D and 3D ultrasonic image of urinary bladder in a dog

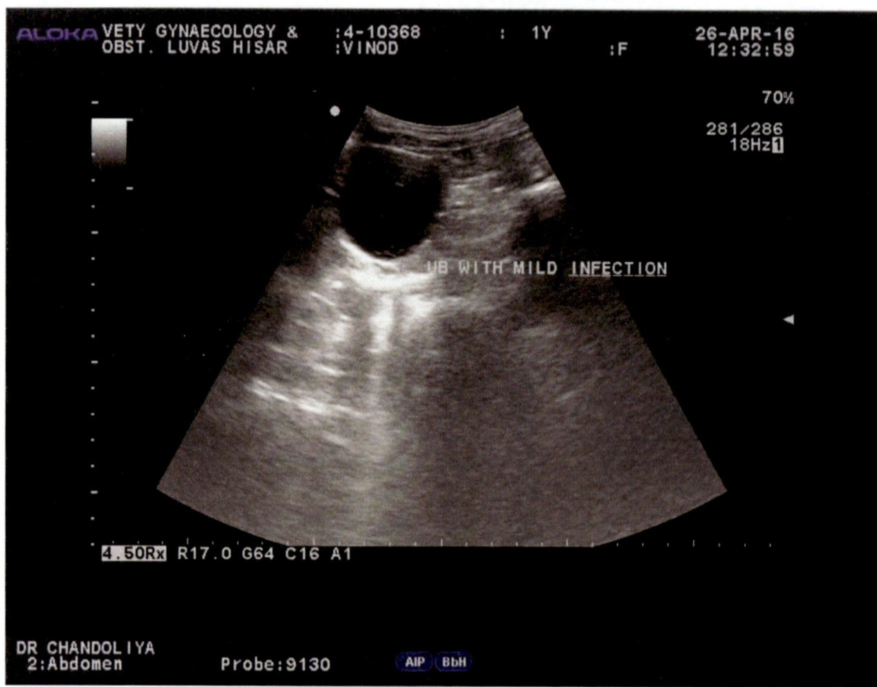

Fig. 197: is showing echoic reflection of mild infection in the urinary bladder in a dog.

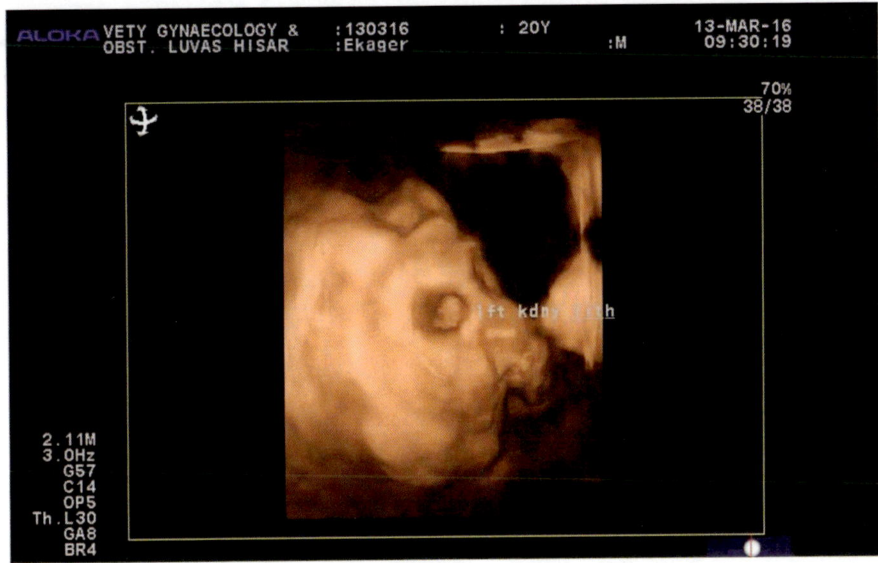

Fig. 198: is showing Kidney stone in 3D USG

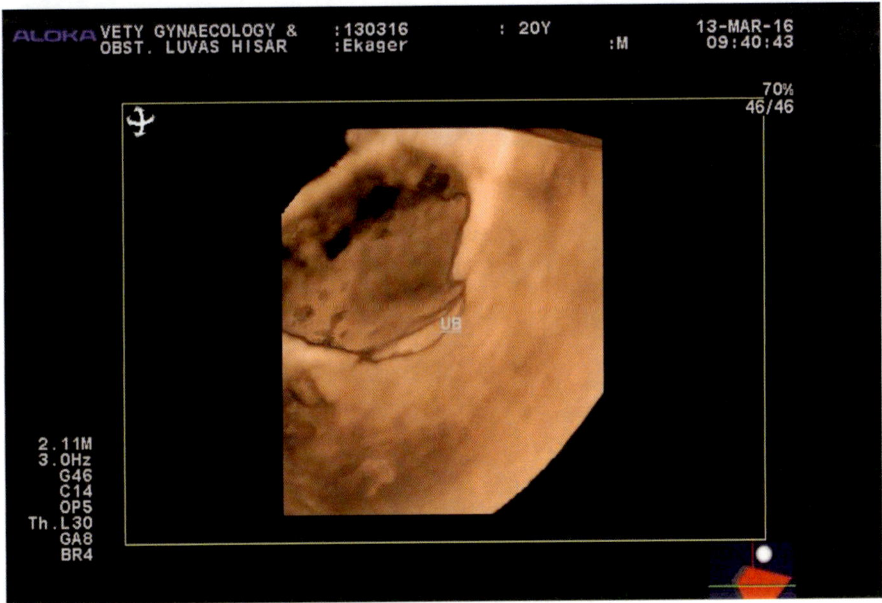

Fig. 199: is showing cross surface view of urinary bladder in 3D scanning.

Scanning for Kidney

Position of probe for left kidney: Photograph-38

Kidney should be imaged in lateral recumbency. For scaning of left kidney, the transducer need to be placed below the transverse process of first to third lumbar vertebrae.

Position of probe for right kidney: Photograph-39

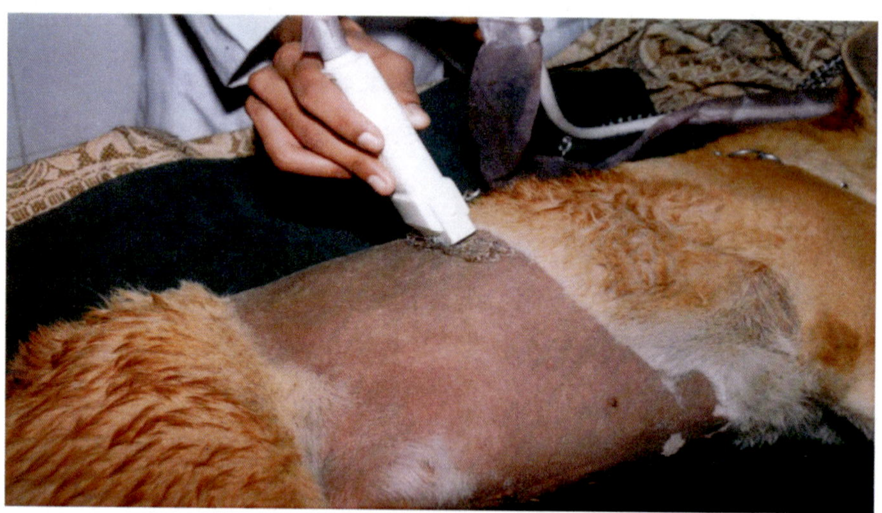

Images of Kidney Showing Nephroliths and Shapes of Kidney Due to Focus of Beam from Different Angles of Transducer

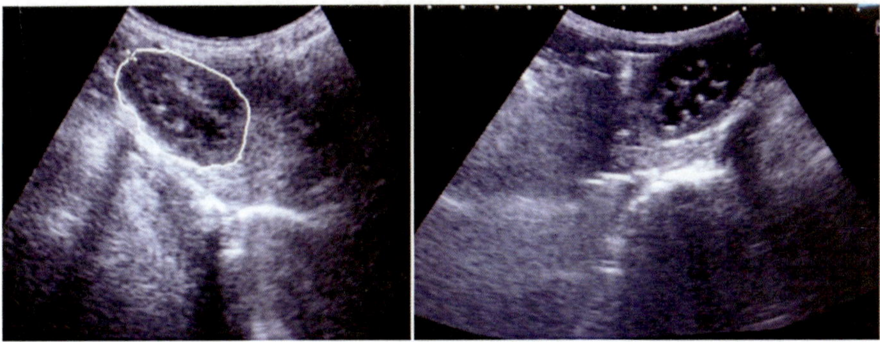

Fig. 200 & 201: are showing kidneys showing ecogenic renal diverticulae and well defined Cortex-Medulla junction of pup of 120 days **(left)** and 180 days **(right)**.

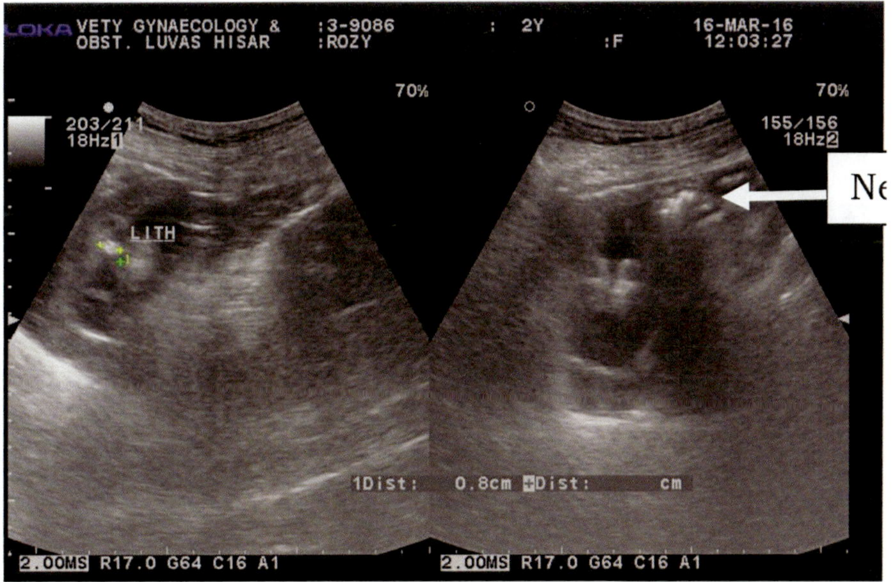

Fig. 202: is showing nephrolith of 0.8 cm in a dog

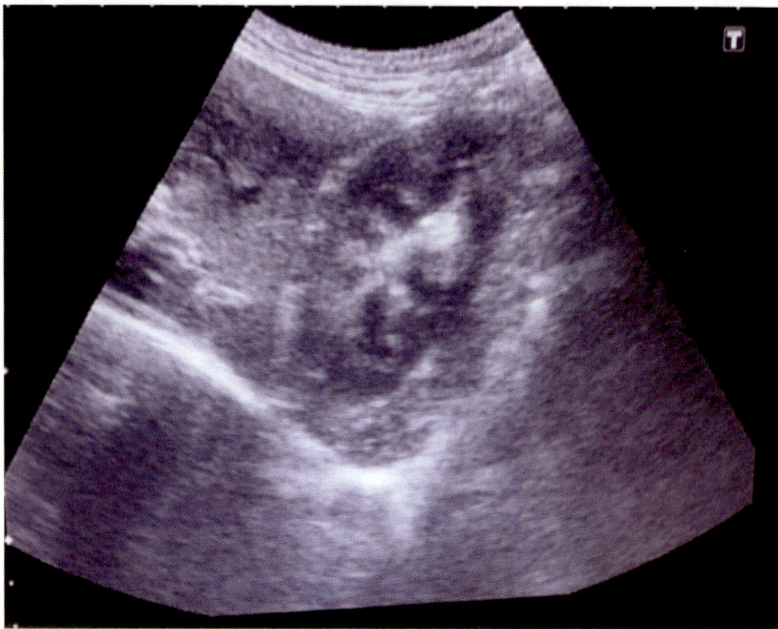

Fig. 203: is showing nephritis in a dog: increased echogenicity of renal parenchyma and less differentiation of cortex and medulla.

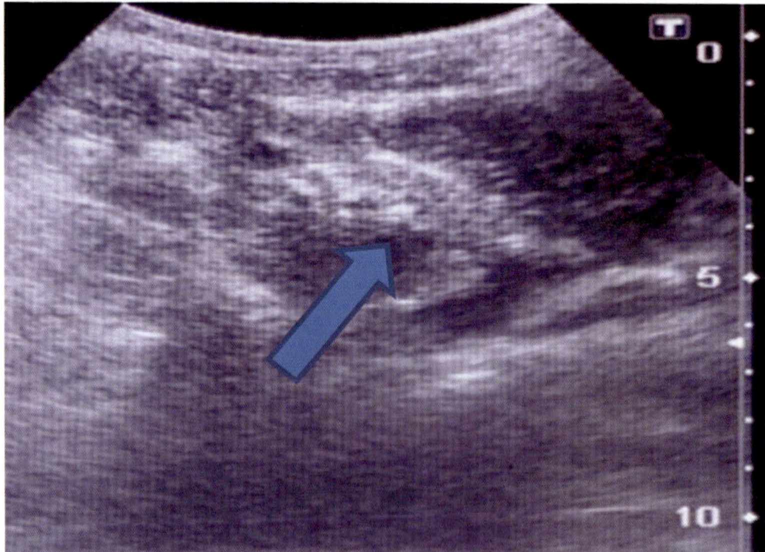

Fig. 204: is showing end stage kidney **(Arrow)**.

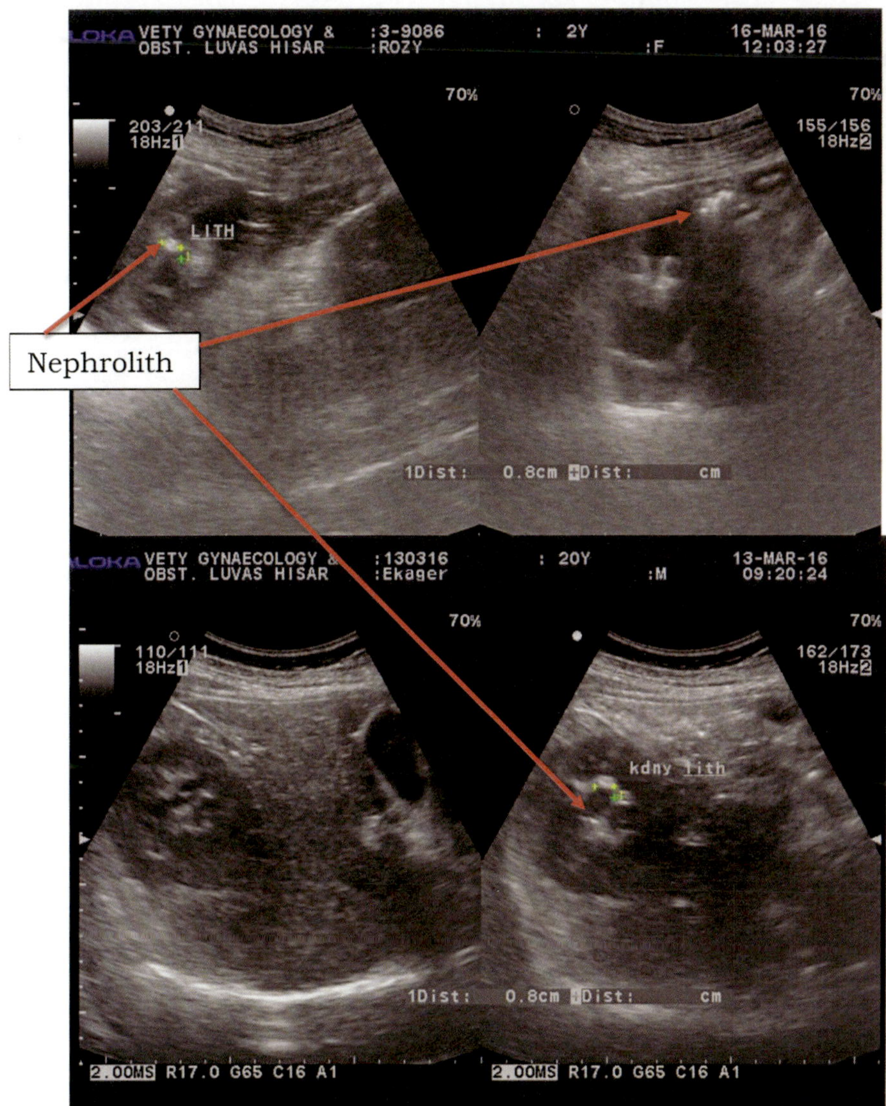

Fig. 205: and 206 are showing nephroliths in the dog.

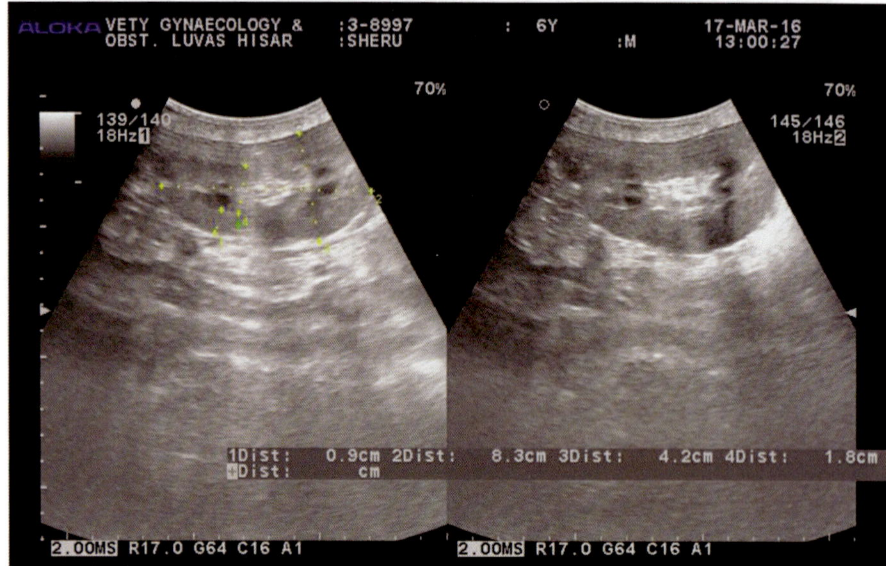

Fig. 207: is showing left kidney in an adult dog measuring (cortex 0.9 cm, medulla 1.8 cm; length 8.3cm & width 4.3cm).

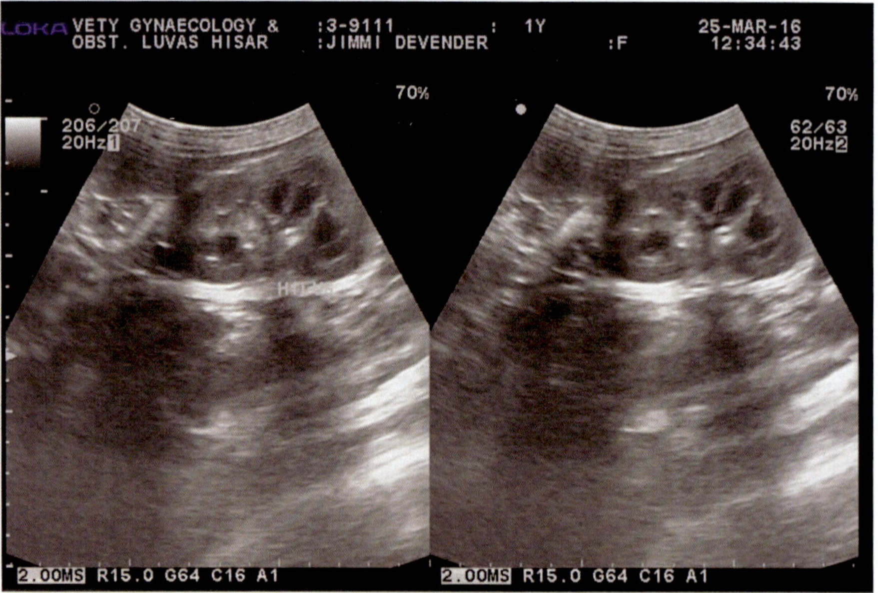

Fig. 208: showing left kidney of dog

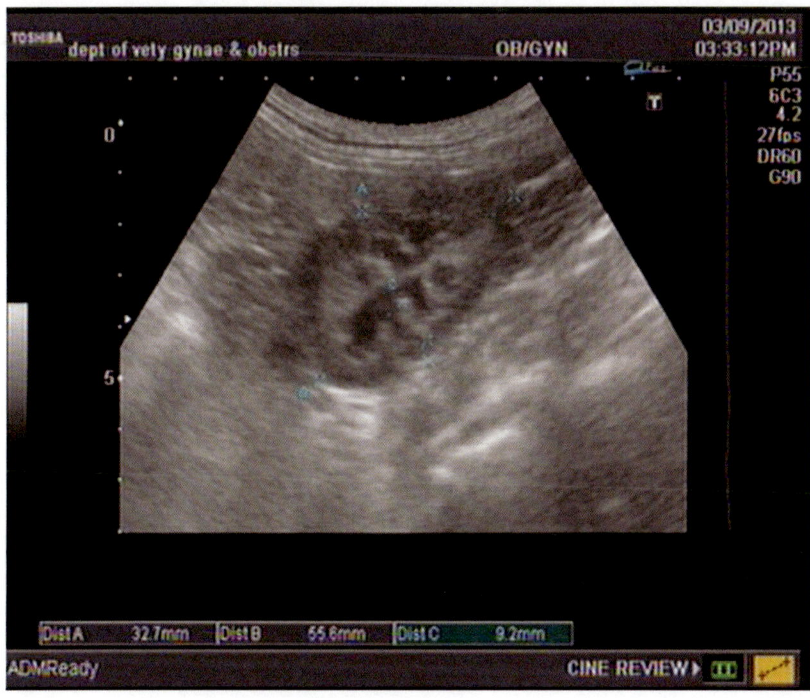

Fig. 209: is showing size of right Kidney in a dog

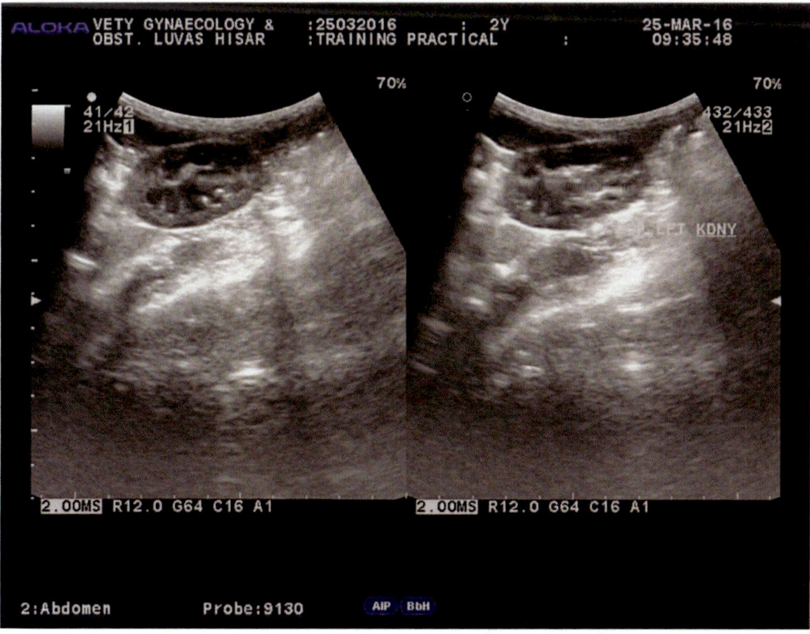

Fig. 210: is showing the Kidney of a pup

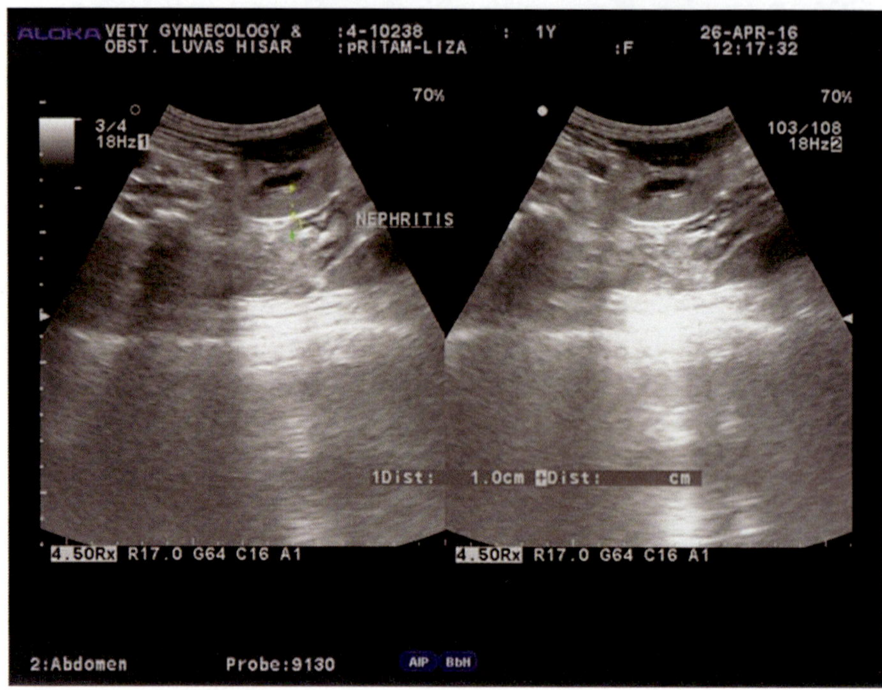

Fig. 211: is showing Nephritis as thickening of capsule is seen

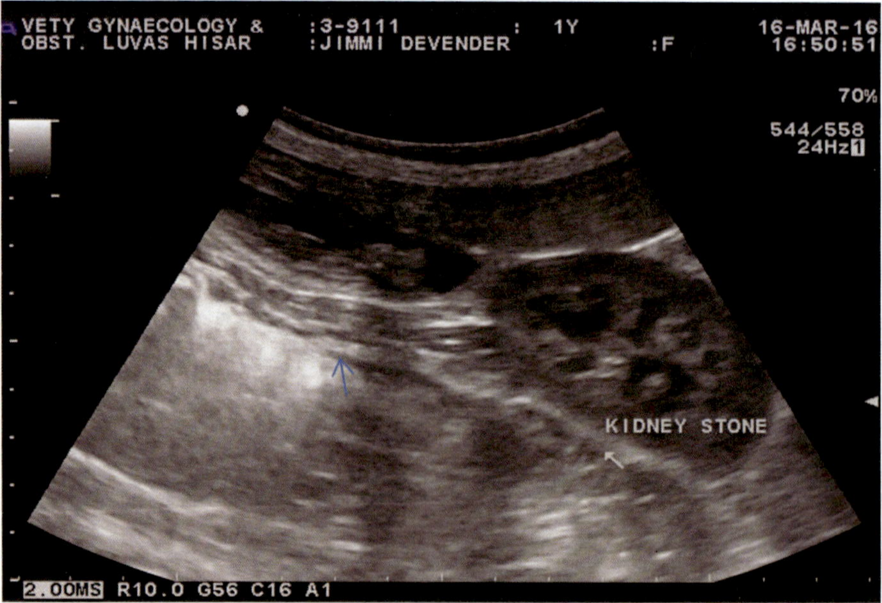

Fig. 212: is showing nephroliths in dog **(top)** & pup **(middle)** and shape of left kidney (bottom).

Probe Position for imaging gall bladder and liver

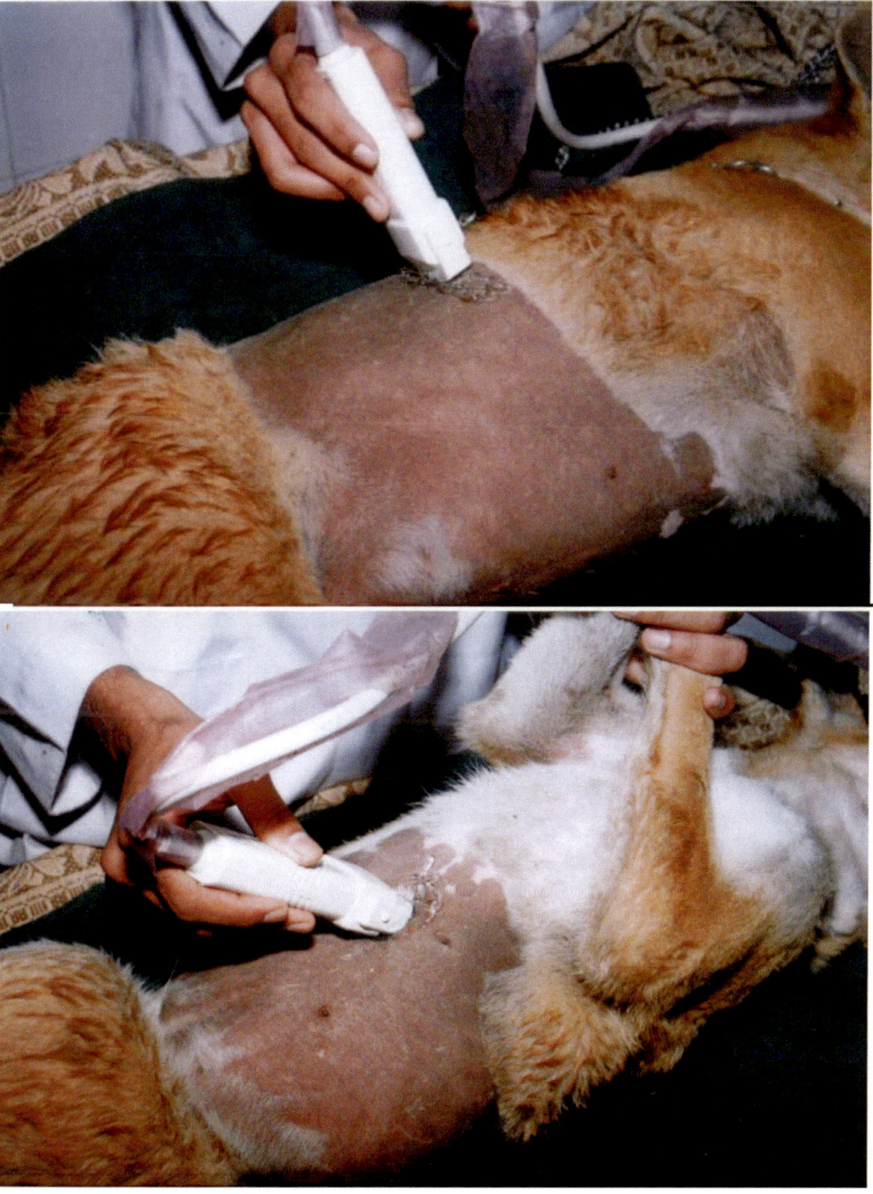

Photographs 40 & 41: are showing position of probe for right lobe of liver **(Upper)** and medial lobe **(Lower)**. The gall bladder can also be visualized from medial side position. Right kidney and right lobe of live can be imaged from this position

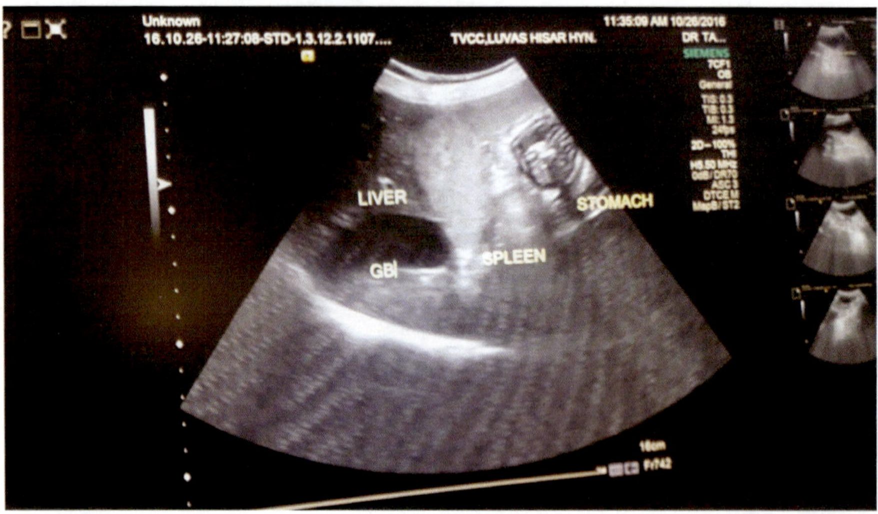

Fig. 213: is showing relative position of abdominal organs in a dog

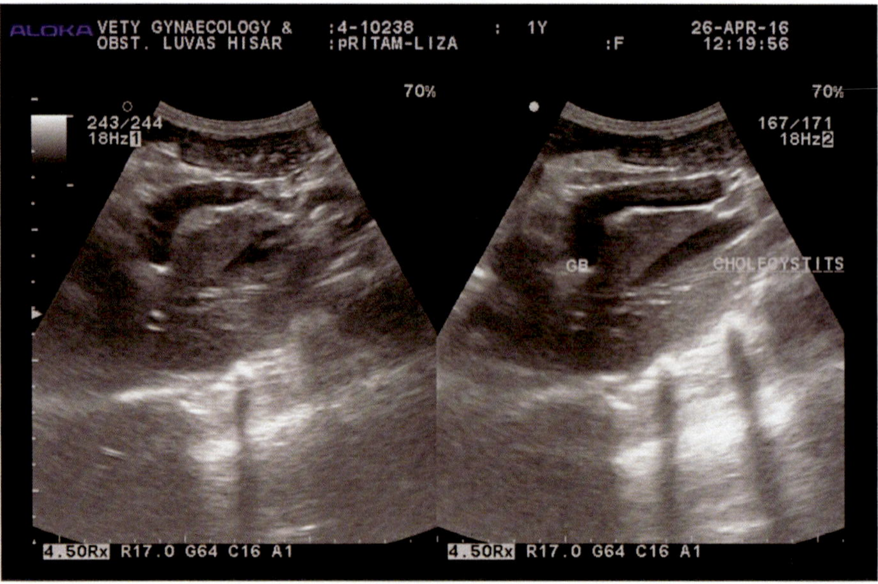

Fig. 214: is showing echoicwall of gall bladder in a dog indicating cholecystitis

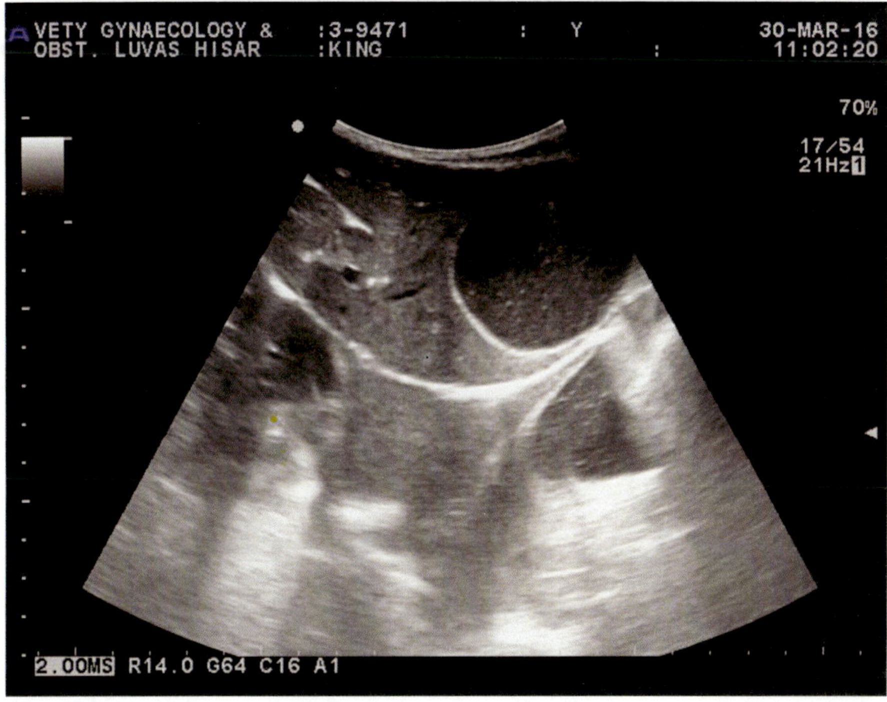

Fig. 215: is showing sludge in GB of dog

48 day

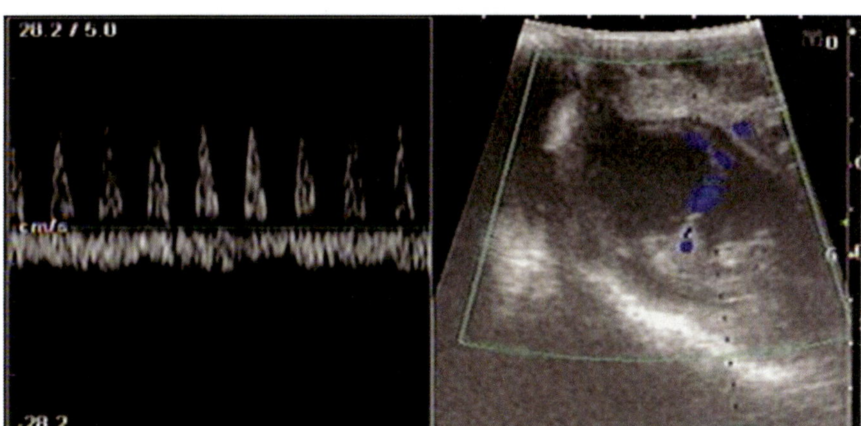

Fig. 216: showing urinary bladder with echoic wall in a dog.

Images of Gall Bladder in Pathophysiological Conditions

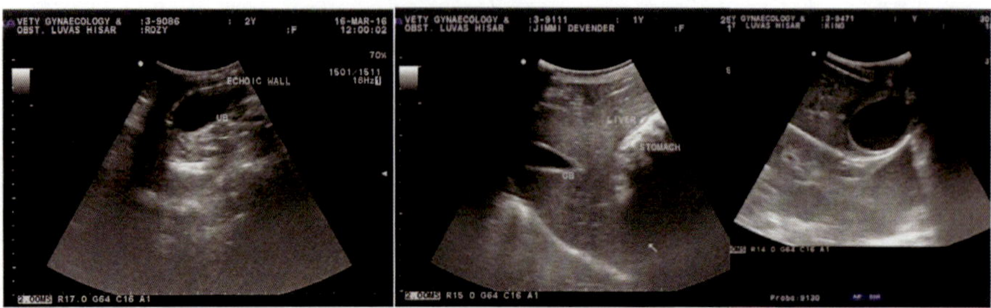

Fig. 217 & 218: are showing Gall bladders of dog with echoic wall (cholecystitis)

Probe Position for imaging stomach: Photograph-42

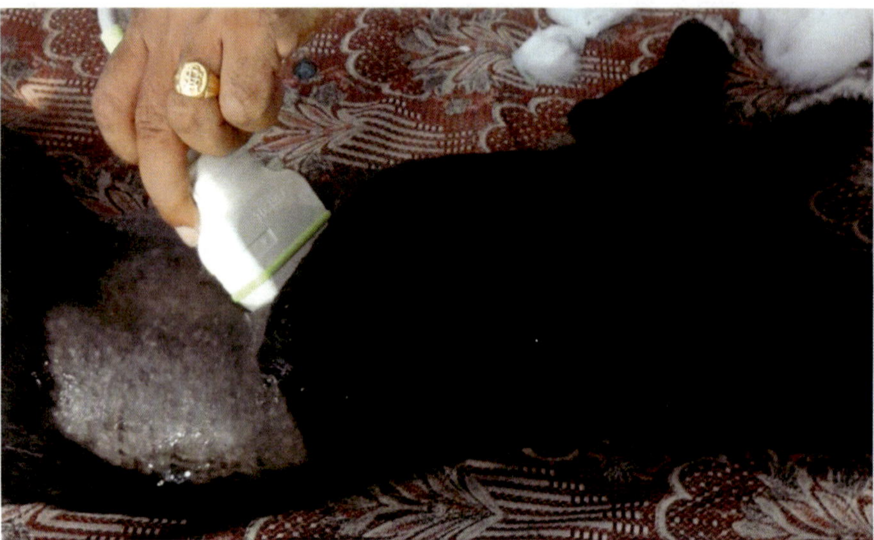

Stomach shoud be imaged in left lateral decumbency keeping in view the topographic position of the stomach just below the left rib cage

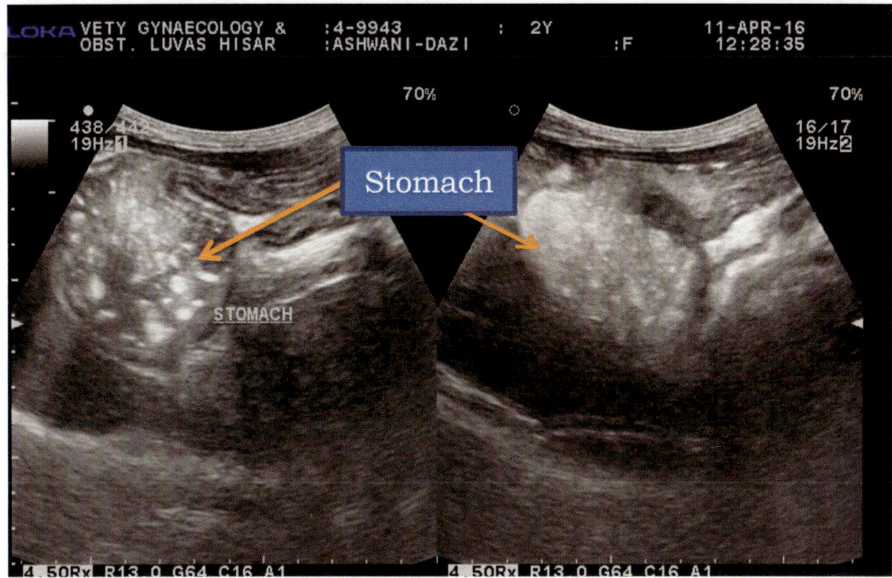

Fig. 219: is showing stomach with contents in a dog

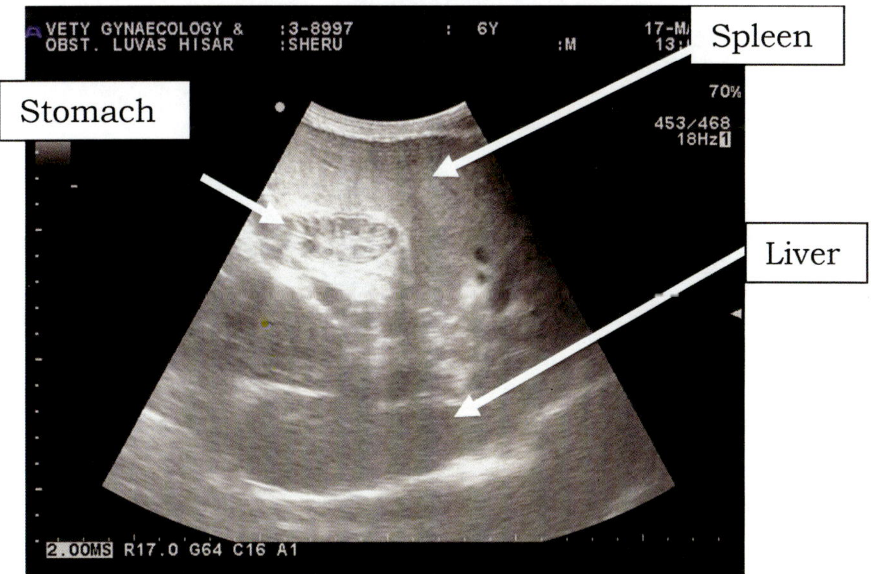

Fig. 220: is showing contracted stomach, spleenand left lobe of liver in a dog

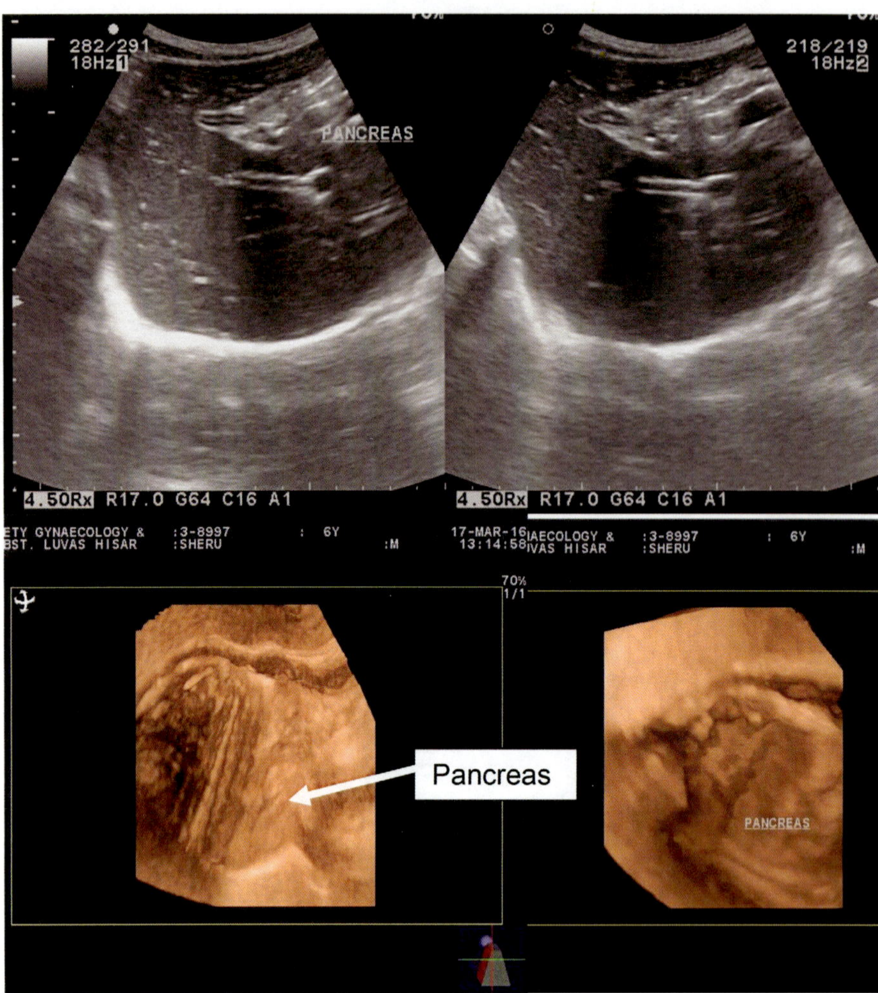

Fig. 221 & 222: are showing 2D **(above)** and 3D **(below)** view of pancreas of a dog

Scanning of intestines for abnormalities

Probe Position for scanning intestine for intusucception, abnormal contents etc. by moving down over abdomen.

Photograph- 43: is showing linear flat surface transducer that should be used to scan intesines.

Intestines can be scanned by using linear transducer, generally used in human USG. If it is not available then linear rectal probe may be used for this purpose. This may be essential in cases where dog engulf some abnormal materials. In one case the dog was scanned to trace a marriage ring. This was located in the intestine as shown in the ultrasonic image below.

Scanning Position for Imaging Uterus and its Contents

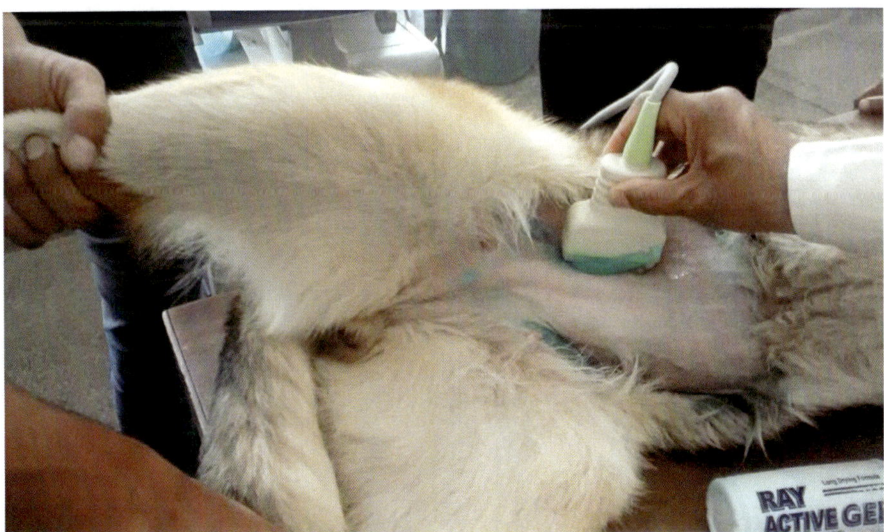

Photograph 44: is showing scanning position for imaging uterus in early pregnancy. The scanner should be moved along line of mammary glands on both side of linea alba.

Images of Uterus in Different Pathophysiological States in A Dog.

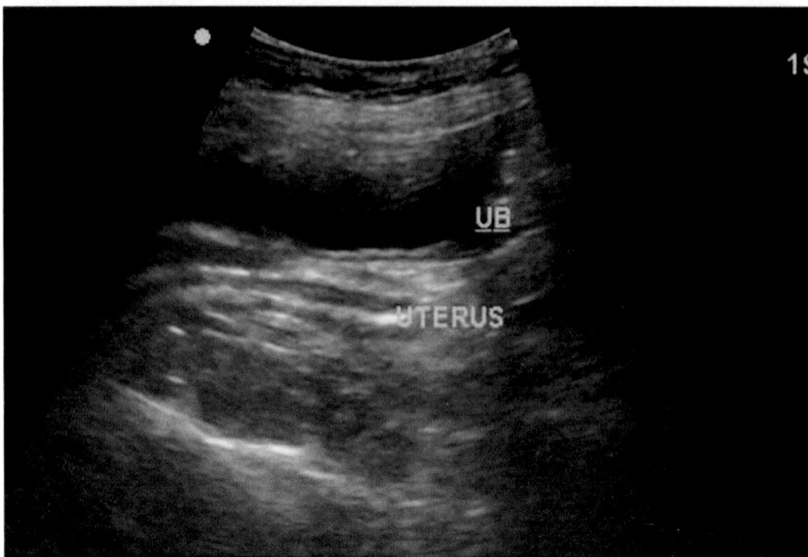

Fig. 223: is showing relative position of non-pregnant uterine horn and urinary bladder in a bitch

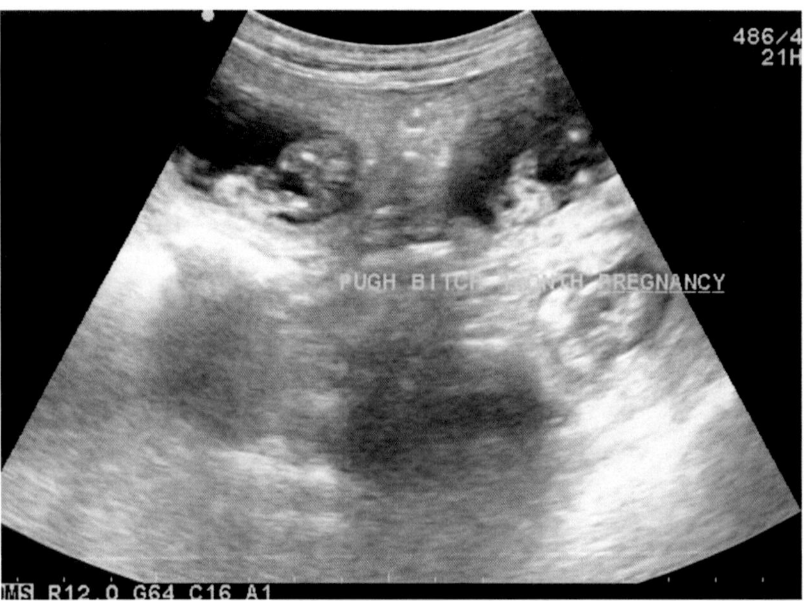

Fig. 224: is showing three pups in a bitch of Pug breed.

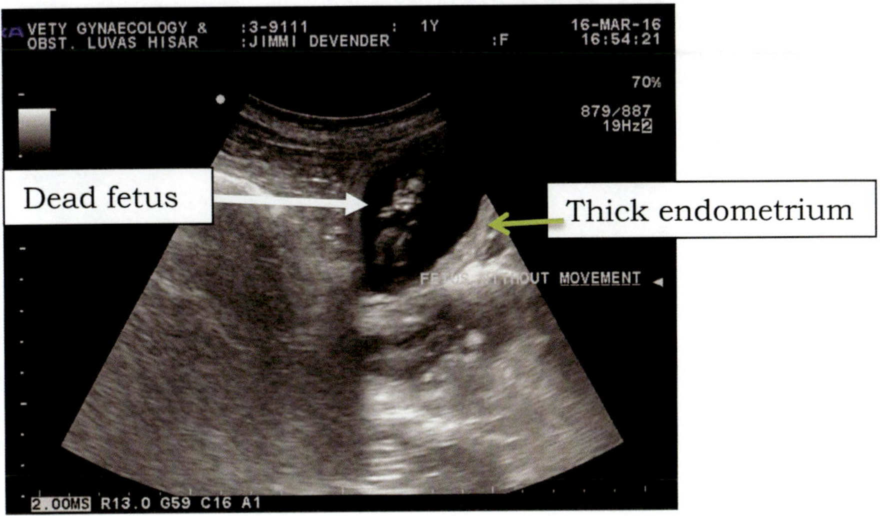

Fig. 225: is showing a dead fetus in a bitch mated 40 days ago.

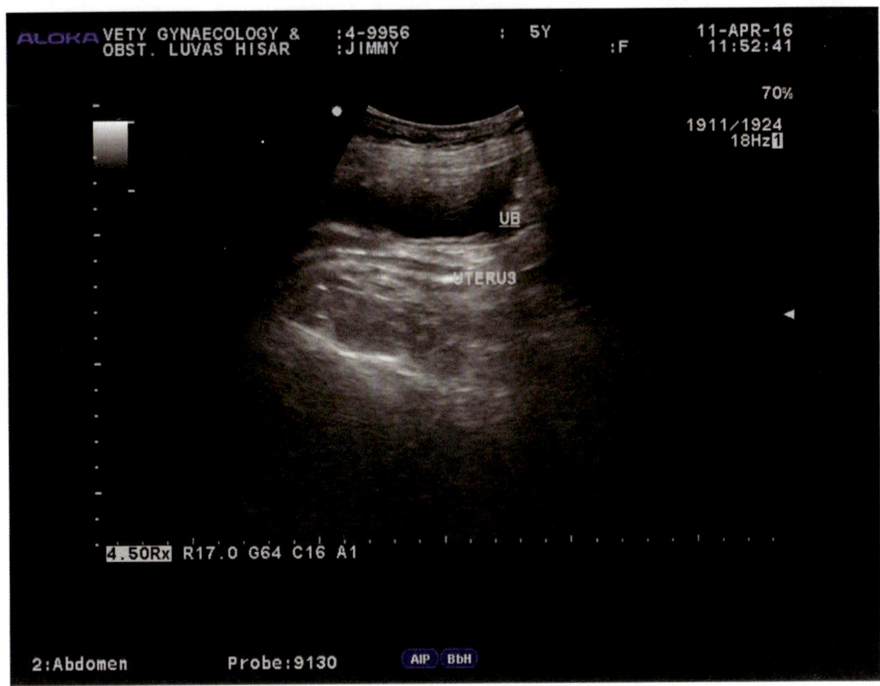

Fig. 226: is showing thickening of uterine wall in a non-pregnant bitch

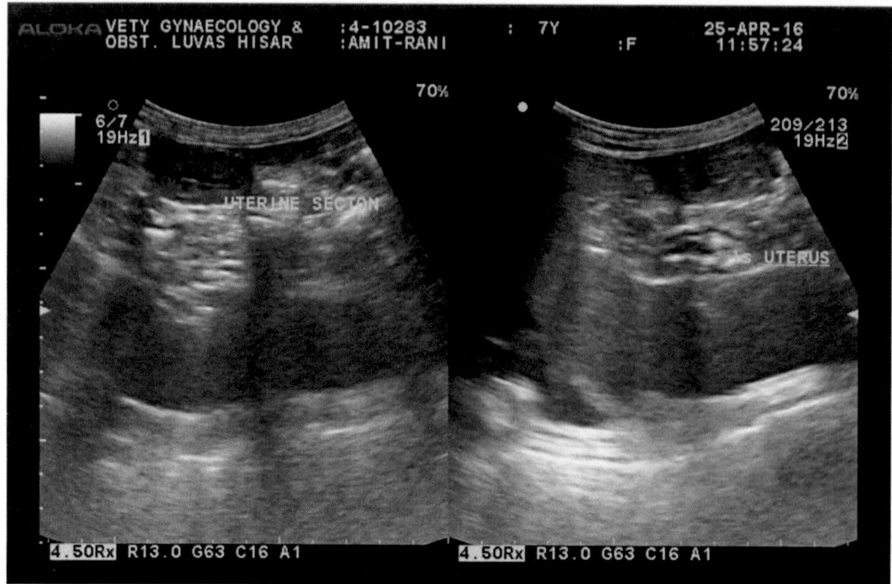

Fig. 227: is showing section of uterus with small quantity of fluid

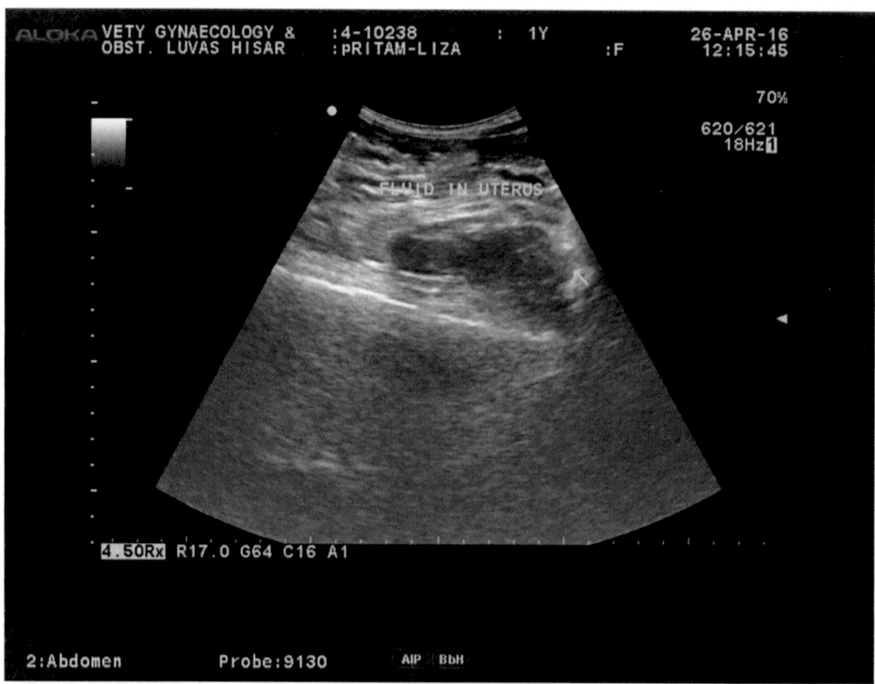

Fig. 228: is showing presence of fluid in uterus

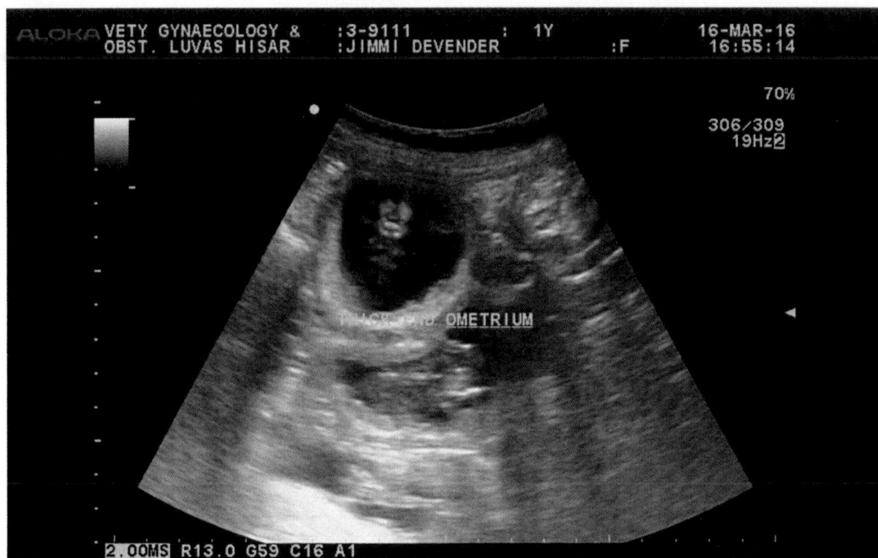

Fig. 229: is showing a dead foetus without movement in uterus with thickening of endometrium

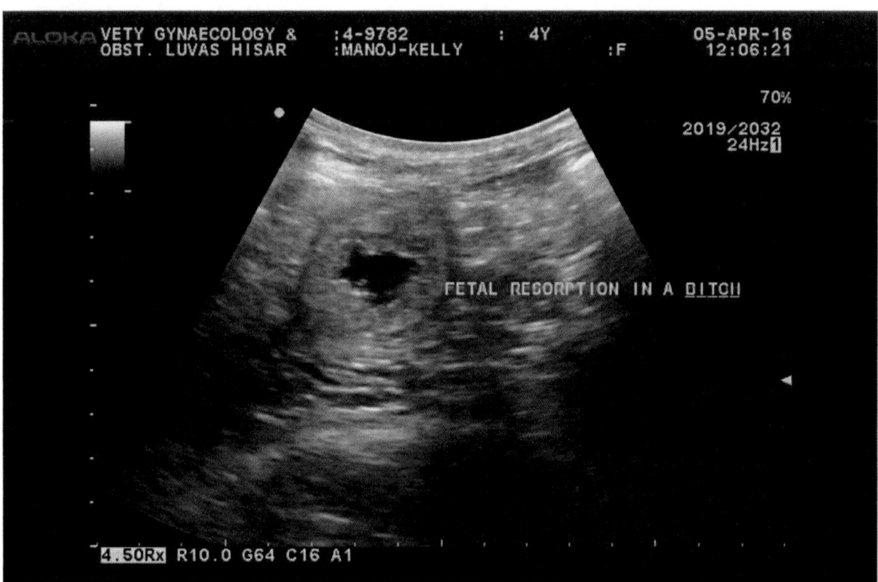

Fig. 230: is showing fluid with echoic contents in uterus with thickening of uterine walls in the bitch indicating foetal absorption

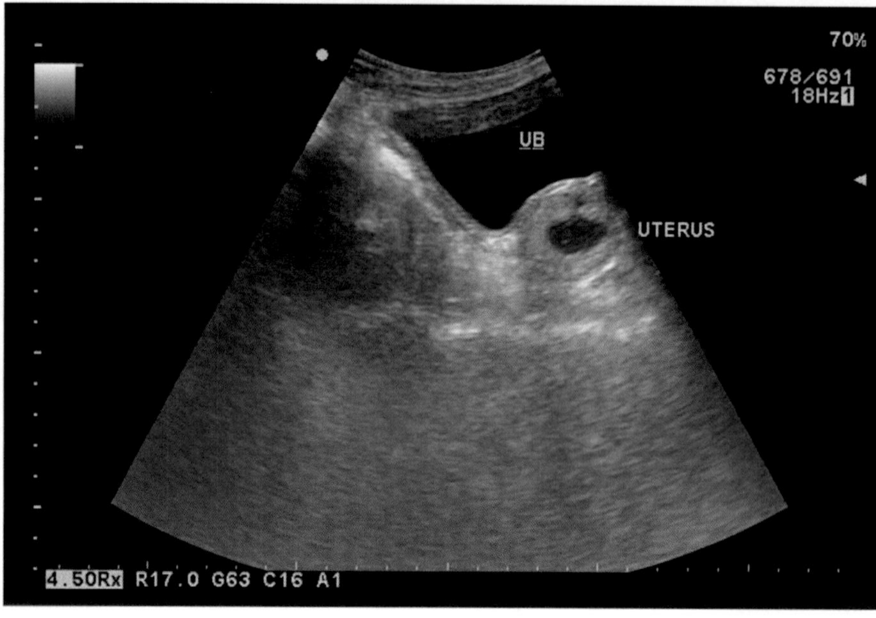

Fig. 231: is showing section of uterus with fluid in the vicinity of urinary bladder

30

Position of Probe in Buffalo for Imaging Different Organs

Photograph- 45: is showing position of probe for scanning advance pregnancy and changes in uterine contents during late abnormal pregnancy

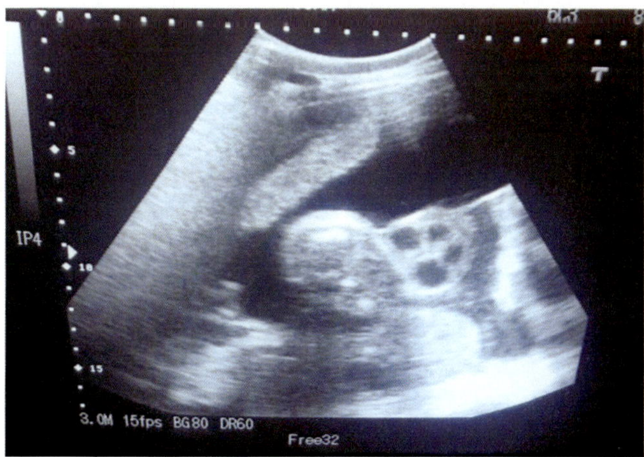

Fig. 232: is showing four vessels of umbilicus of fetus from transaabdominal ultrasonography

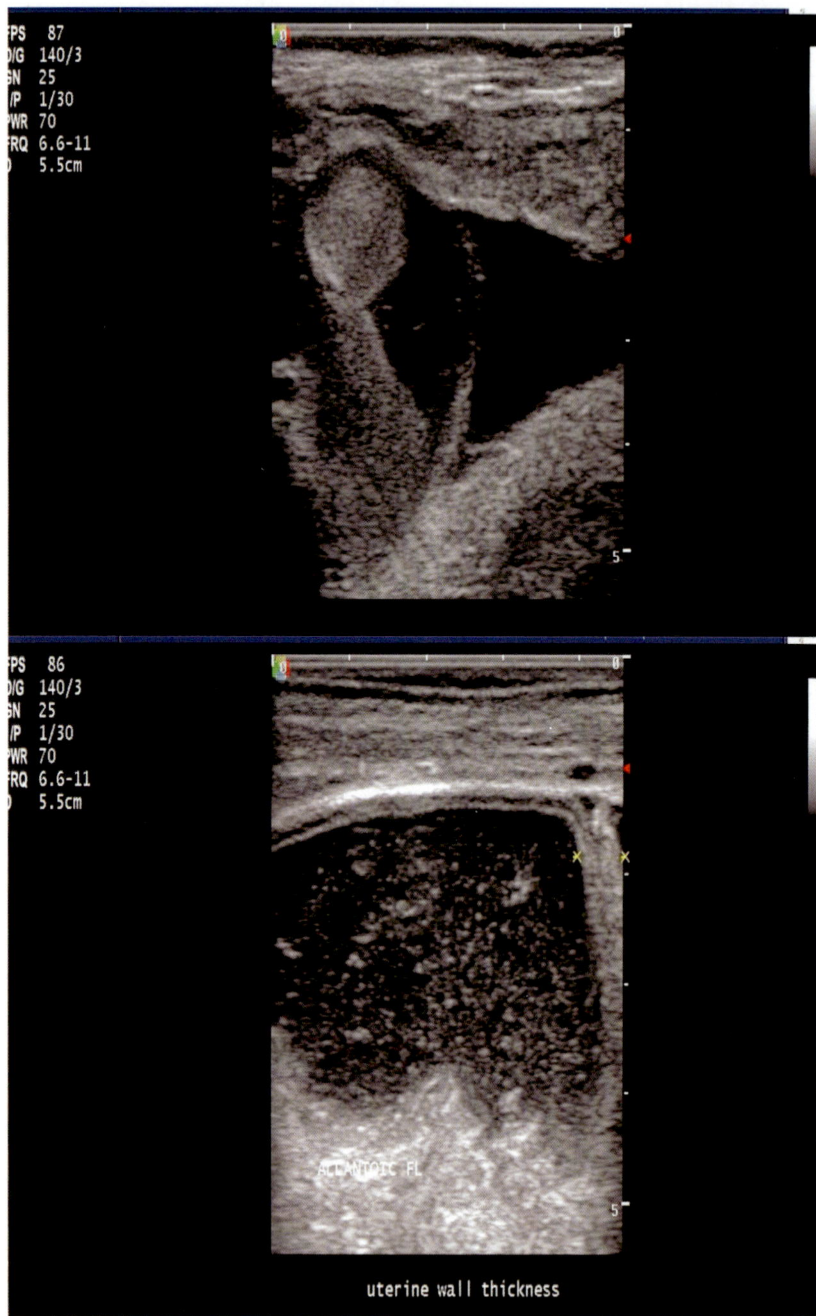

Fig. 233 & 234: are showing normal fetal fluid with normal cotyledons **(above)** and abnormal uterine contents **(below)** in buffaloes.

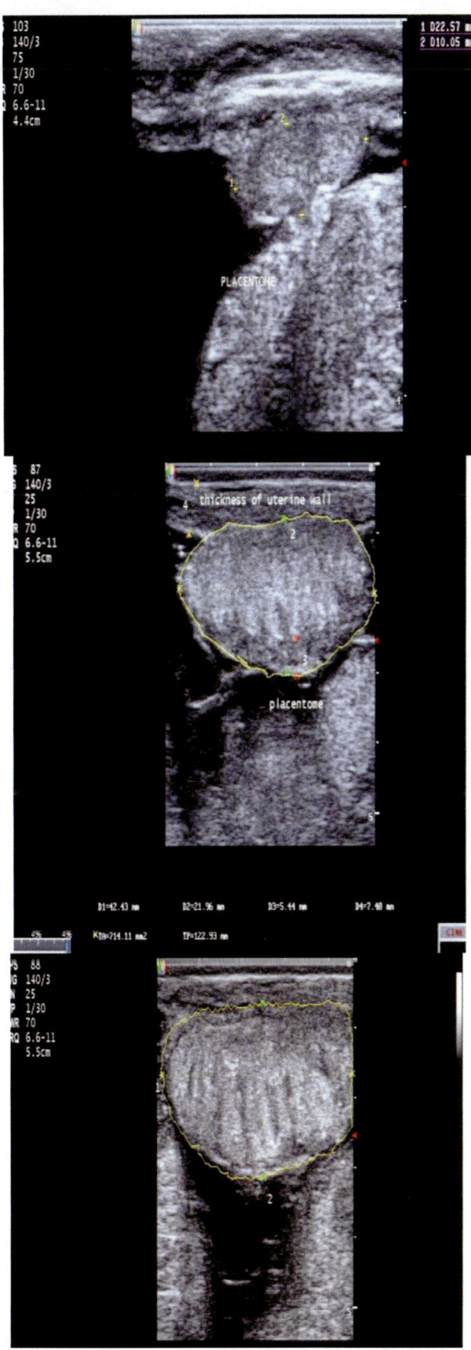

Fig. 235 & 236: are showing normal compact cotyledon with normal outline **(top)**, enlarged cotyledon with abnormal margins **(middle)** and fluid filled internal pocket in cotyledon **(bottom)** in a buffalo suffering from uterine torsion.

Scanned Area of Kidney, Liver, Spleen and Heart in Buffalo

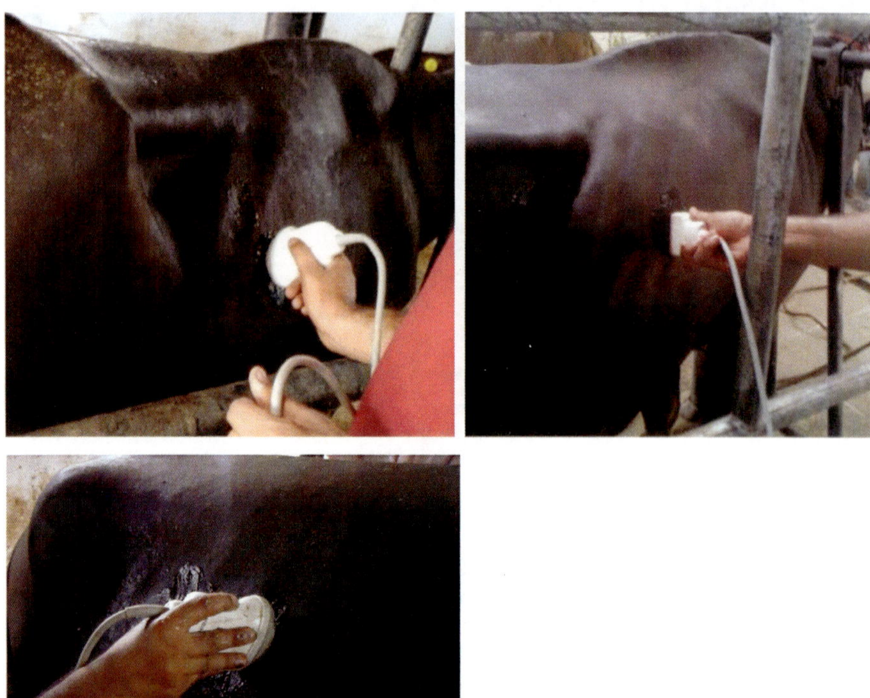

Photograph 46- 48: are showing site and probe positioning for transcutaneous ultrasonography of liver in buffaloes. The right thoracic area between 12th to 8th intercostals space is good for trans- thoracic ultrasonography of liver.

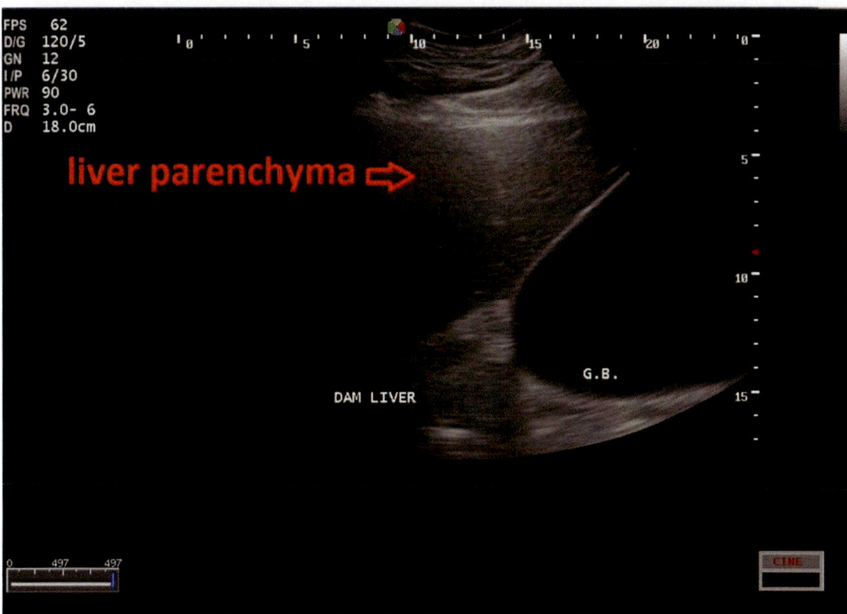

Fig. 237: is showing transabdominal sonographic image of liver along with gall bladder (GB) in normal advanced pregnant buffalo with 2D convex transducer having frequency 3.0-6.0 MHz (examined from 11th ICS).

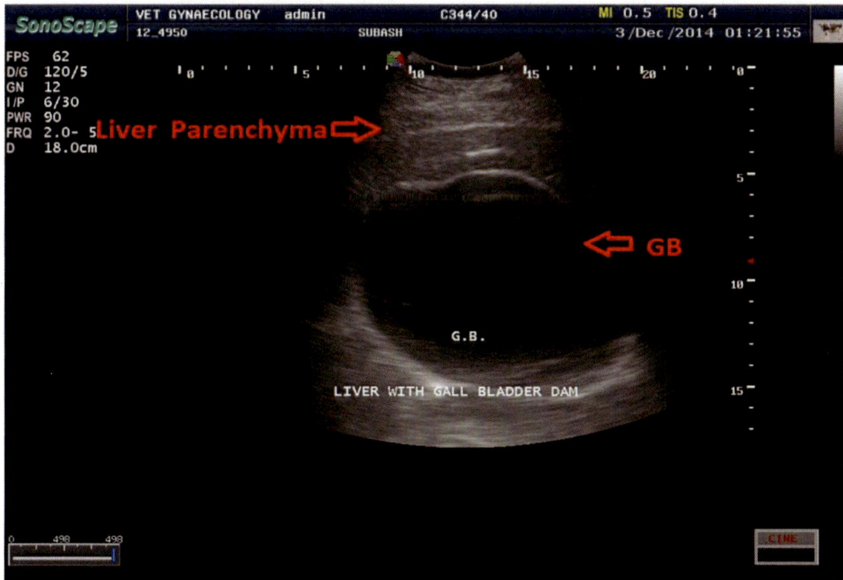

Fig. 238: is showing Transabdominal sonographic image of liver along with gall bladder (GB) in full term buffalo having right side post cervical torsion since 12 hours with 2D convex transducer having frequency 2.0-5.0 MHz (examined from 11th ICS).

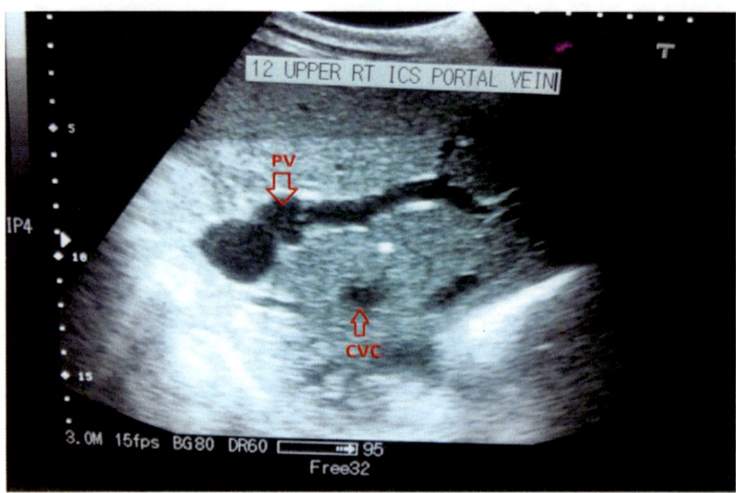

Fig. 239: is showing transabdominal sonographic image of liver with major blood vessels named as portal vein (PV) and caudal vena cava (CVC) in normal advance pregnant buffalo with 2D convex transducer having frequency 3.0 MHz (examined from 12th ICS).

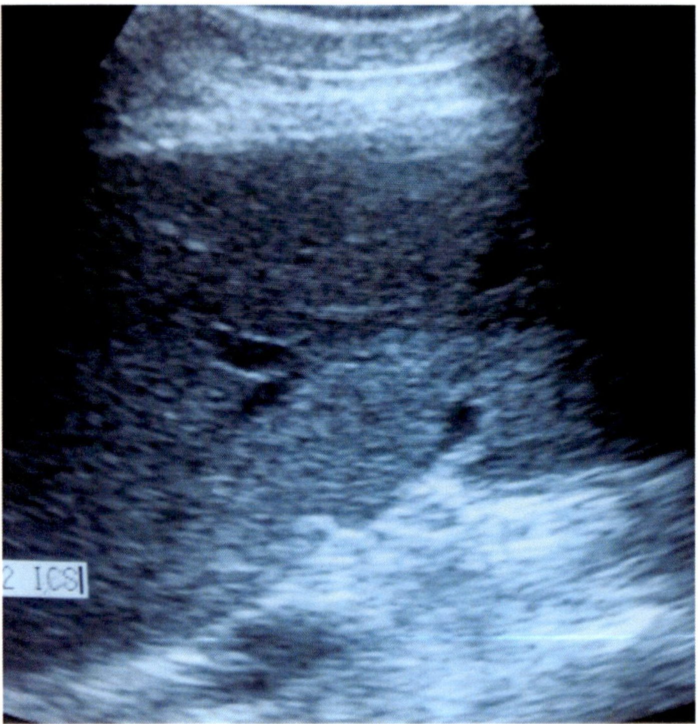

Fig. 240: Echotexture of liver in Murrah buffalo. Homogenous echoic pattern is normal ultrasonogram

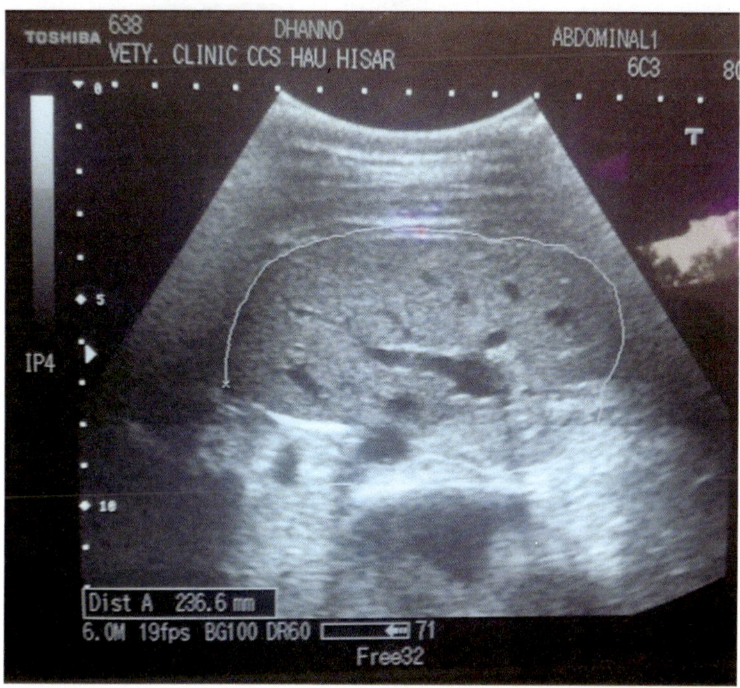

Fig. 241: is showing hepatic vein and portal triad in a liver in a buffalo

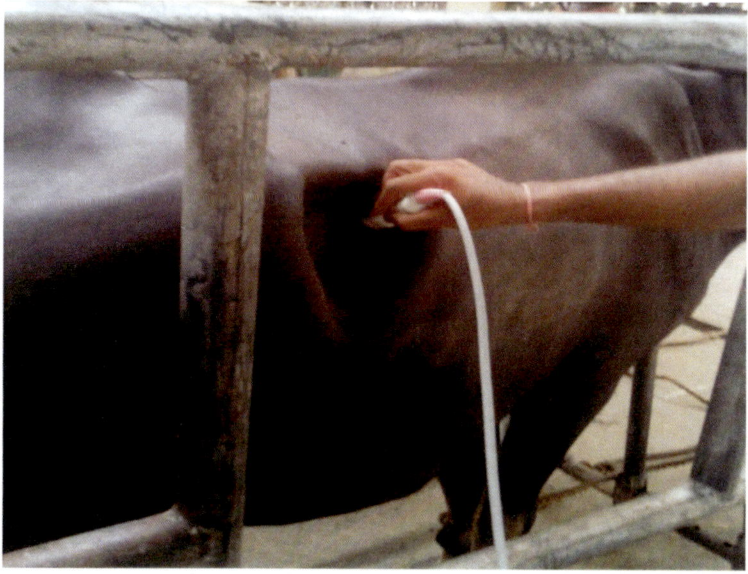

Photograph- 49 is showing trans-abdominal ultrasonography for right kidney from right paralumbar fossa. The transducer is directed perpendicular to right paralumbar fossa while, the scanning of left kidney is problematic from left paralumbar fossa due to a hindrance by rumen.

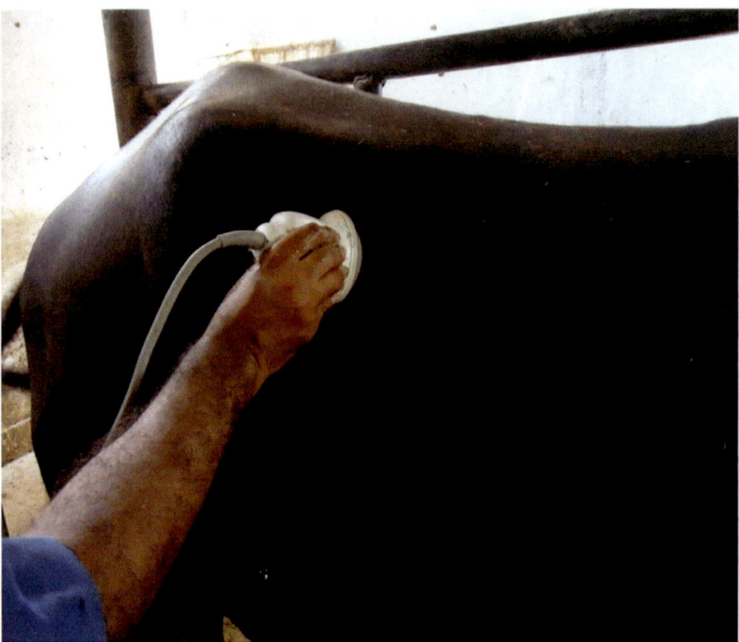

Photograph-45 are showing sites and probe positioning for transabdominal ultrasonography of right kidney from right Para- lumbar fossa in buffaloes.

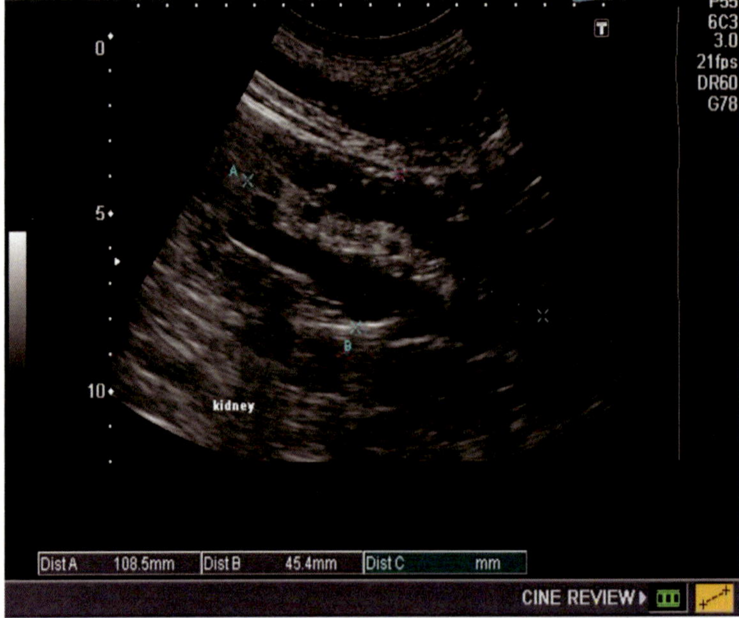

Fig. 242: is showing measurement of length of right kidney in a buffalo.

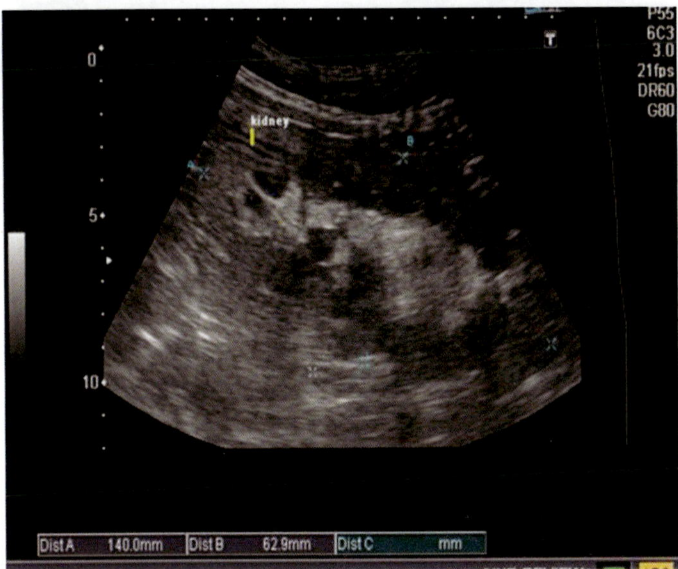

Fig. 243: is showing width of right kidney in a buffalo

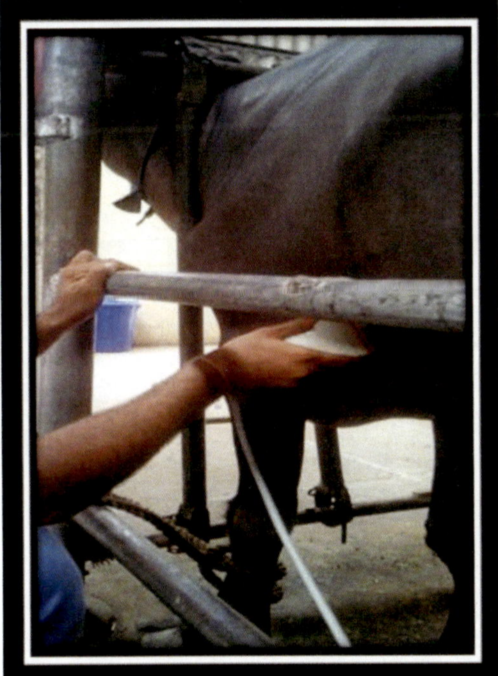

Photograph- 50: is showing scanning position of heart in a buffalo. The third and fourth intercostal spaces in both the left and right sides of the thorax is shaved for ultrasonography of heart.

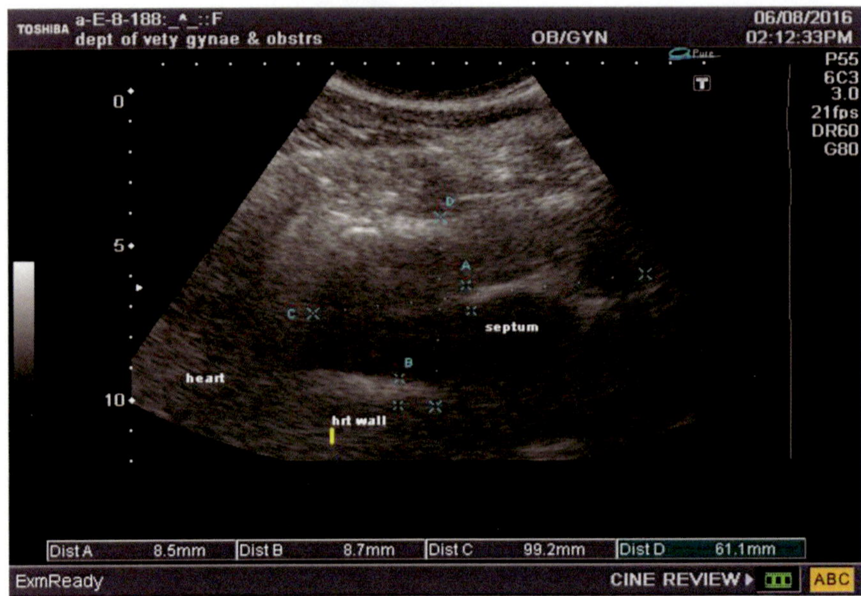

Fig. 244: is showing image of heart of buffalo, measuring ventricular wall thickness.

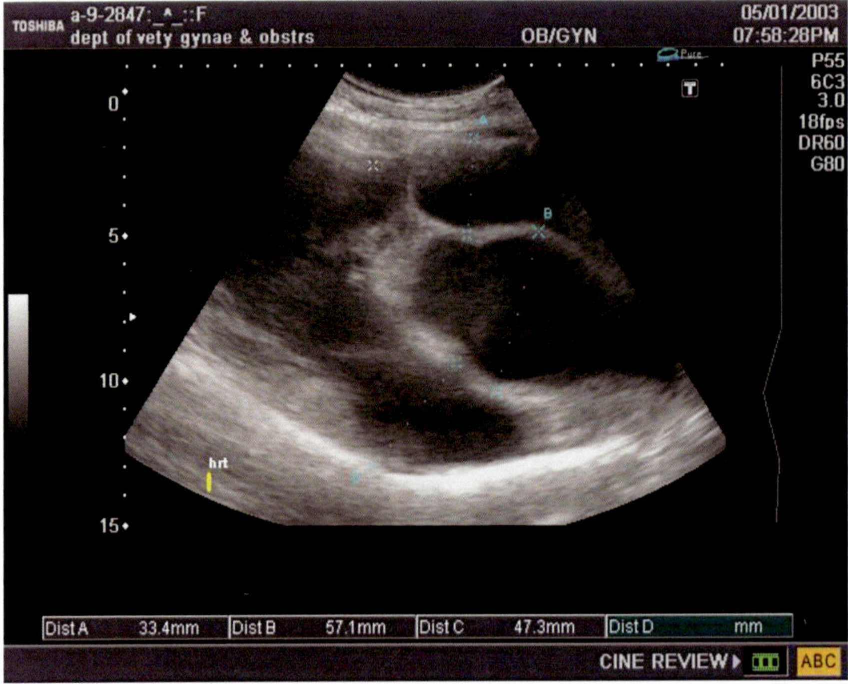

Fig. 245: is showing measurement of atrial diameter in a buffalo.

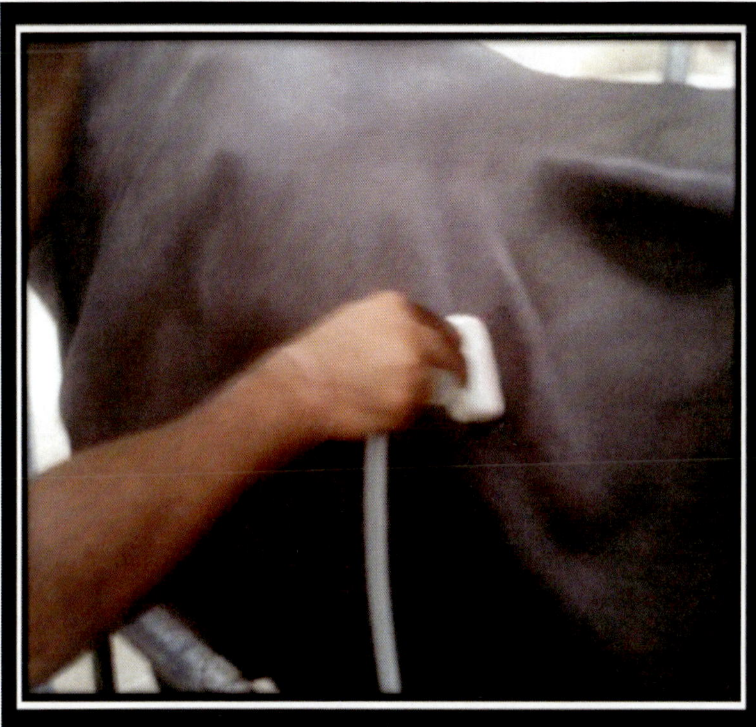

Photograph- 51: is showing trans-thoracic ultrasonography for spleen in a buffalo. The area on left side between 11th to 9th intercostals space is shaved for spleen.

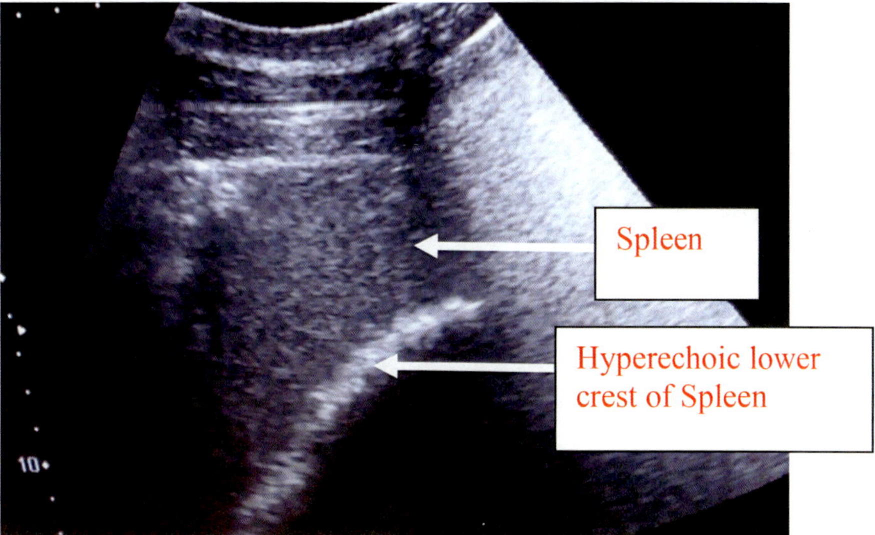

Fig.-246: is showing spleen of a Murrah buffalo, using 4.2 MHz convex transducer

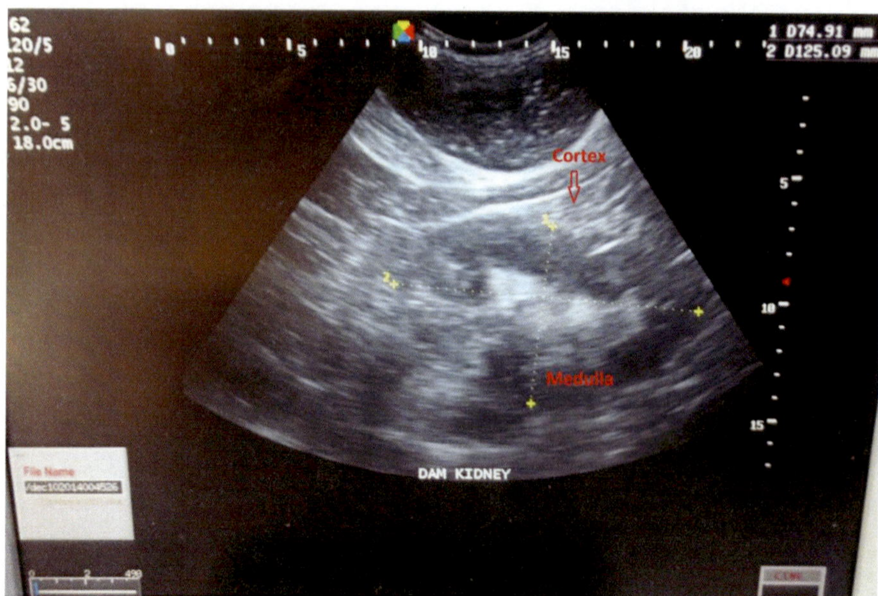

Fig. 247: is showing transcutaneous sonographic image of right kidney in normal advanced pregnant buffalo with 2D convex transducer having frequency 2.0-5.0 MHz (examined from right paralumbar fossa).

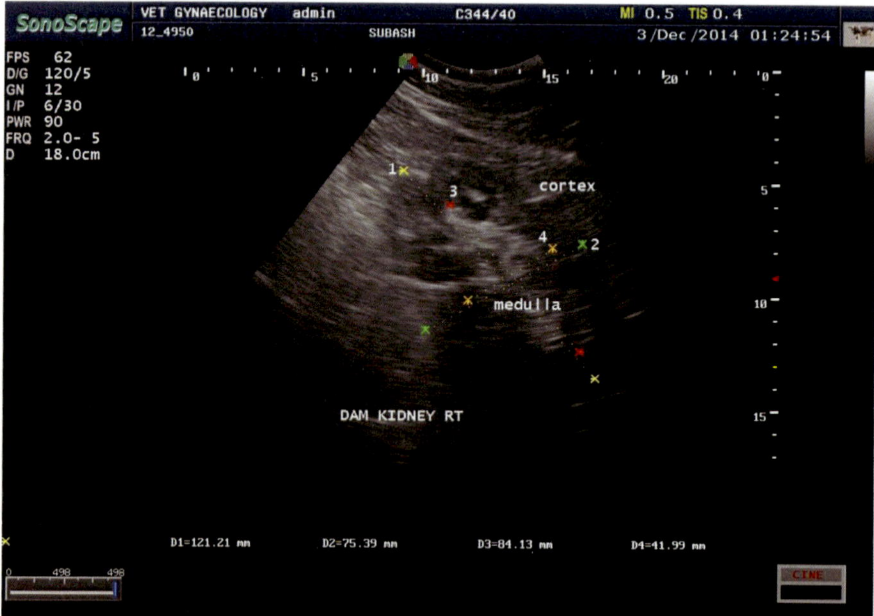

Fig. 248: is showing transcutaneous sonographic image of right kidney in uterine torsion affected buffalo with 2D convex transducer having frequency 2.0-5.0 MHz (examined from right paralumbar fossa).

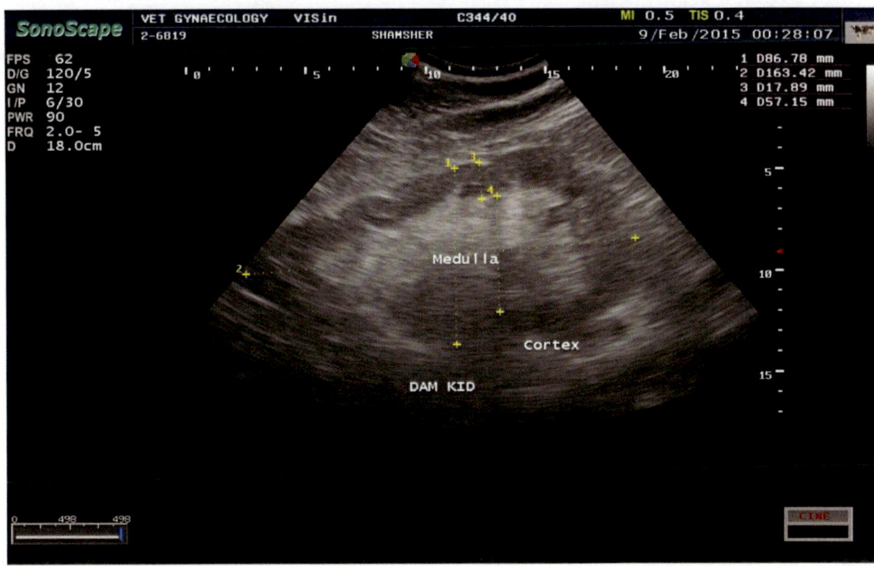

Fig. 249: is showing transcutaneous sonographic image of right kidney in normal advanced pregnant buffalo whose dimensions were measured ultrasonographically with 2D convex transducer having frequency 2.0- 5.0 MHz (examined from right paralumbar fossa).

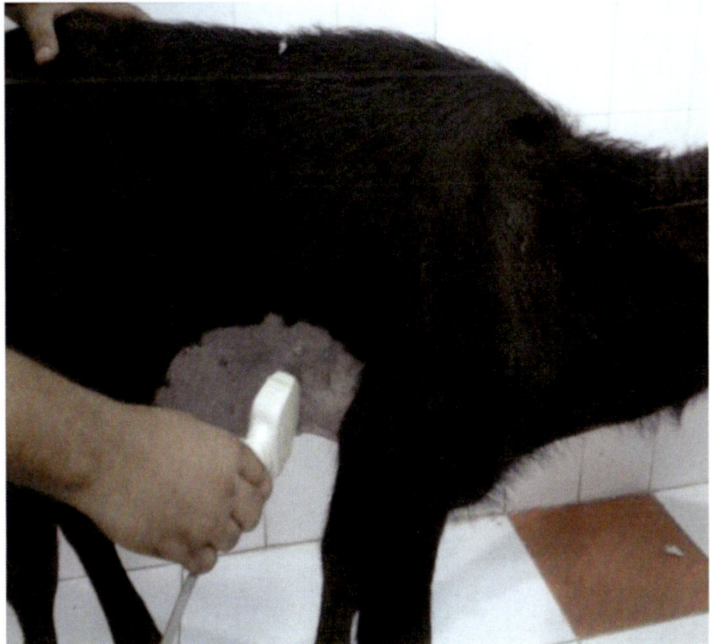

Photograph 52: is showing position of transducer for scanning omasum in a buffalo Calf.

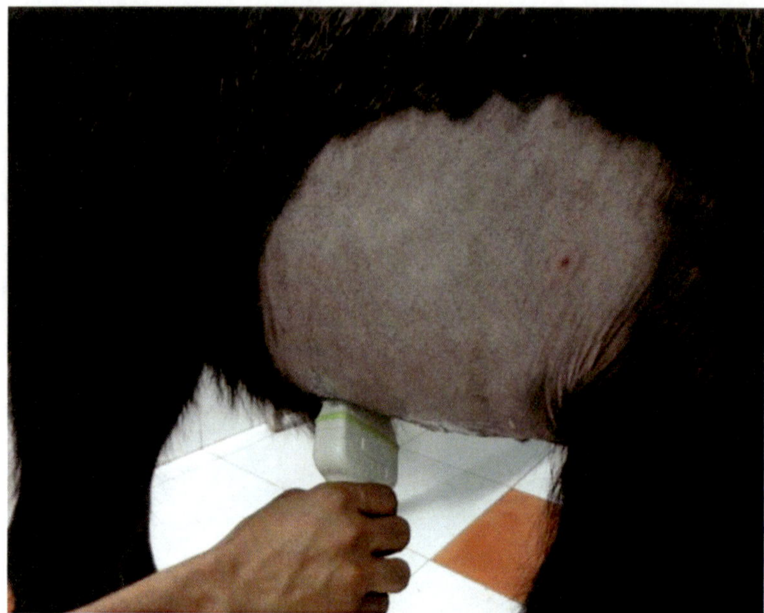

Photograph 53: is showing position of convex probe of 3.0 MHz frequency for abomasum image in a buffalo calf.

31

Ultrasound Report Writing

Veterinary ultrasonographer has to write report of ultrasonography in different animals. The proper ultrasound report writing is the moral duty of the vet ultrasonographer. The accuracy of diagnosis must be ensured. The poor resolution/quality of machine may lead to erroneous diagnosis. Therefore, veterinarian must ensure good quality of machine and proper diagnosis. The report might be different in different animals e.g. report writing about uterus will be different in bitch, cows, buffaloes, camel, horse, goat and sheep. However following might be guidelines in writing ultrasound report of uterus. fetus, endometrium

Uterus: Position of uterus, contents in the uterus, shape and size in initial pregnancy and infections, appearance of myometrium, thickening of uterine layers, presence or absence of fluied/pus/masses, left and right horns and their contents. Observations on the images of uterine fluid and its echoicity, membrane appearance in pregnant animals. Special attentions should be paid to endometrium.

Endometrium: Appearance and thickness; any growth on the surface of endometrium. Images of retained portion of placenta can be ascribed in the cases of retained placenta.

Ovary: Ovaries are scanned to assess cyclic structures, size, shape, its Position/ location and internal echo pattern of follicles. This echo pattern will indicate whether follicle is normal, cystic or with any other abnormalities of follclular fluid. Ultrasound reports also include shape and size of corpus luteum. In cows and buffaloes, presence of cavity, looseness/compactness of luteal cells are also mentioned.

Fallopian Tubes: Many a times, it is difficult to get proper image of fallopian tubes. If these are visible, then mention that these were visible. Mention their diameter, presence of fluid or mass or any other abnormalities, if observed.

Fetus Status Report: Livebility of fetus and status of fetal fluid is important to note and mention in the report. There are several parameters of fetus that must be measured/observed and put these into your report. First of all see, if fetus is present or not. If present fetal heartbeat is present or not. Count heartbeat and report it. There are inbuilt programs in good machines to help in counting of heartbeat. Report on status of fetal fluid; whether anechoic or echoic with abnormal reflective particles. Observations on amniotic and alloantoic membranes. Mention the diameters of these membranes, if it has been taken. In bitch, swine, goat and sheep and other species, mention number of foetuses present. In early pregnancy, measure size of fetus in length, width and circumference (before limb appearance). Later on take measurement on crown rump length, head circumference, trans-cerebellar diameter also referred as biparietal diameter (BPD). In small animals, take measurements of abdominal circumference, femur length to correlate and calculate fetal age. These must be mentioned while detailing about fetus in ultrasound report. Observations on placenta should also be mentioned e.g. presence and observations on cotyledons, their size and shape, compactness in bovine, caprine and ovine. In abnormalities of pregnancy e.g. in fetal death, looseness in internal composition of cotyledons are seen and should be mentioned in ultrasound report.

Report writing on Cervix

Size of cervix or diameters or circumference of Cervix. Measure these diameters and report. Check condition of cervix and report any kink cervix, double cervix, mention location of cervix. Whether shape is normal or there are echoic reflection i.e. cervicitis. One should also check folds carefully and report about any abnormal structure, if observed.

Report on KIDNEYS

Mention size i.e. dimentions and circumference; observations about Shape i.e. normal or abnormal. Observations about characteristics of cortex and medulla, collecting system, main and intra-renal arteries and veins, renal capsule, outline of kidney normal or any abnormality.

Urinary Bladder

Appearance of wall, normal wavy, any breakage or any other abnormalities of wall. Mention about appearance of contents whether normal or abnormal with reflective particles or pusy contents.

Ureters

It is difficult to take images of ureter. Good ultrasound machine are helpful in taking images of ureters. Assessment of the presence/absence of dilatation.

Liver

In small animals, sixe and contour and reflection characteristics should be mentioned. Since size is large in farm animals, therefore one should mention about its contour and kind of beam reflection. All segments of liver should be assessed and their characteristics. Characteristics of all segments should be mentioned. Appearance of intrahepatic vessels and ducts viz portal triad should be mentioned. Walls of vessels shoule be carefully observed and evaluated change in dimentions or wall thickness. Hepatitis is commonly observed, however, setting of machine should be proper, otherwise abnormal gain setting may show normal liver as hepatitis case. Porta hepatis and adjacent area, portal venous, hepatic venous and arterial systems should be evaluated but these requied good USG machines.

Gall Bladder

Mention about Size, Shape, Contour, ultrasound characteristics of the wall and the nature of contents. Cholicystitis and sludge in gall bladder are common in dogs.

Spleen

Size, Shape, Contour, Ultrasound characteristics including the hilum, Assessment of splenic vein blood flow, Presence/absence of collaterals. Splenomegaly is seen in dogs.

Prostate: In dogs Shape and size is mentioned. In bull its dimensions are easy to take and mention.

Seminal Vesicles: In bull grape shape seminal vesicles, also refered as vesicular gland is irregular in shape is easily scanned. The round shape of seminal vesicle in this species is indicative of its inflammation. Its circumference and dimensions can be measured and contents can be assessed.

Bulbourethral Gland: Since it is paired gland but covered by urethralis muscle in bull, therefore, this is taken as combined one gland. Take its measurements for USG report.

Gastro-Intestinal Tract: wall thickness, its contents, diameter of lumen, motility and presence/absence of masses are reported.

Diaphragm: Contour, Movement, Presence of adjacent fluid, Presence of masses and Presence of lobulations are reported.

Musculo-Skeltal Examination

Tendon: Examine insertion point for tearing, evidence of free fluid/bursitis, Thickening associated with inflammation, Calcification and partial or complete tears

Pancreas: Size, Shape, Contour, Ultrasound characteristics of head, body, tail and Diameter of main duct are reported.

Ligaments: Appearance of falciform, Ligamentum teres and Venosum are reported in human. Same may be reported in Vet ultrasound report.

Aorta: diameter, course and branches including the bifurcation appearance of its walls, lumen and para-aortic regions are reported.

Other Structures: where relevant include:

Omentum, Muscles, Abdominal wall, Possible hernias, Lymph nodes and sites for potential fluid collection (including upper/ lower abdomen and the thorax)

Following is reports of human for guidance:

Technique: An ultrasound examination of the abdomen was performed. This includes liver, kidney, Gall Bladder, spleen, aorta, inferior vena cava, urinary bladder etc.

Findings

The liver has normal echogenicity. The liver is normal sized.

The gallbladder has a normal appearance without gallstones or sludge. There is no gallbladder wall thickening or pericholecystic fluid. The common bile duct was not visualized.

The pancreas is mostly obscured by gas. A small portion of the head of pancreas is visualized which has a normal appearance.

The aorta has a normal caliber. The aorta is smooth walled. No abnormalities are seen of the inferior vena cava.

The right kidney measures 10.8 cm in length and the left kidney 10.5 cm. No masses, cysts, calculi, or hydronephrosis is seen. There is normal renal cortical echogenicity.

The spleen is somewhat prominent with a maximum diameter of 11.2 cm. There is no ascites.

The urinary bladder is distended with urine and shows normal wall thickness without masses.

The prostate is normal sized with normal echogenicity.

Similar guidance is published in the literature. Two true examples are given from human being that can be taken as format

U/S Opinion

Whole Abdomen

•	Liver	:	Shows normal outlines with normal echo texture. Intra hepatic biliary radicals are not dilated. Hepatic vasculature is normal.
•	G. Bladder	:	Normal distension. No calculus. No evidence of acute or chronic cholecystitis.
•	CBD	:	Normal in size.
•	Spleen	:	Normal in size and reveals normal echo texture.
•	Pancreas	:	Normal in size, shape and echo pattern.
•	Right kidney	:	Show normal outlines. It measures 105 x 43 mm in size. No calculus. No hydronephrosis. No renal mass.
•	Left kidney	:	Show' normal outlines. It measures 119 x 53 mm in size. No calculus. No hydronephrosis. No renal mass.
•	Retroperitoneum	:	No lymphadenopathy.
•	U. Bladder	:	Normal outlines. No calculus. No mass.
•	Prostate	:	Normal echo texture.

Impression : No Sonolog1cal Abnormality Detected.

USG of Upper Abdomen - 3.2.2004

Renal bipolar lengths, RK - 9.96 cm, LK = 6.89 cm.

Renal parenchymal thicknesses. RK = 1.04 cm, LK = 0.923 cm.

The lcft kidncy is smaller compared to the right. The pelvlcaiyceal sy stem of the left kidney is dilated. No calculi are seen tn the right or left kidney.

No focal lesion is seen in the pancreas, spleen, liver and gallbladder.

Multiple diverticula are noted in the bladder. No calculi or mass is seen in the bladder.

Impression

The left kidney is smaller in sixe compared to the nght. Dilatation of the left peivicalyceal system is noted.

Multiple diverticula are noted in the bladder which is in keeping with a neurogenic bladder.

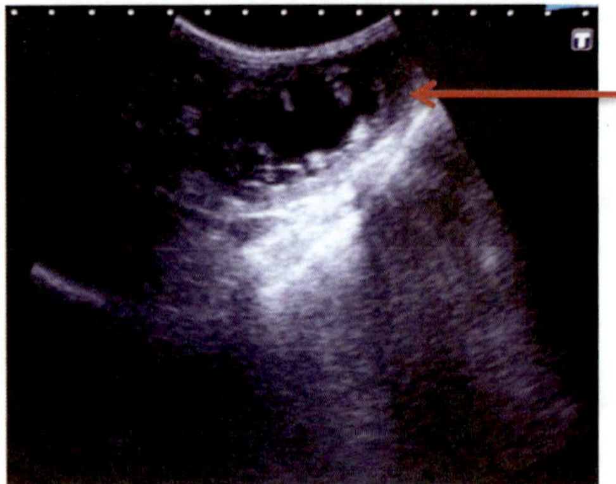

Case 22: Ultrasound image of Hydronephrosis in a dog. Arrow is showing fluid accumulation the kidney.

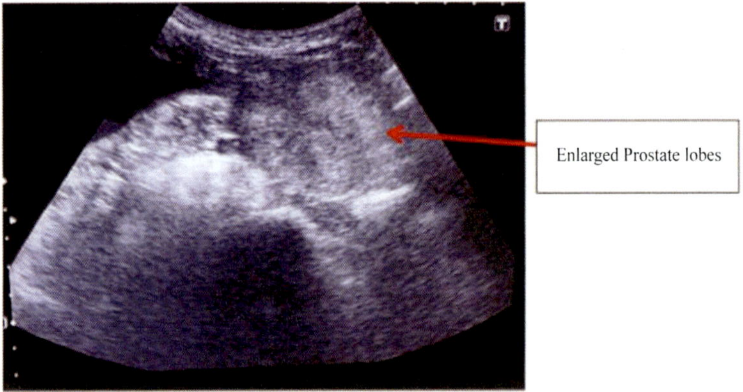

Case 23: A dog was having difficulty in urination. On ultrasonography enlarged prostate was diagnosed.

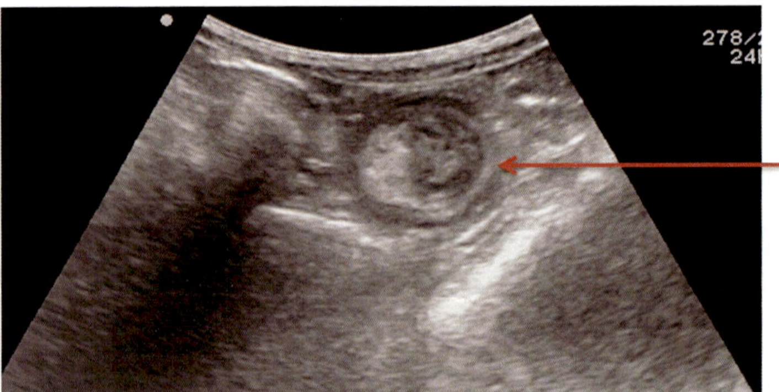

Case 24: A marriage ring of a couple was lost. The ultrasonography showed image of ring (arrow) in the intestine of their dog.

References

Abdel Rahim, S.E., Abdel Rahman, K., Nazier, A.E. (1994) Production and reproduction of one humped camels in the Al-Qasim region, Saudi Arabia. J. Arid. Environ. 26, 53-59.

Aboul-Ela, M.B. (1994) Reproductive performance of one humped camel under traditional management in the United Arab Emirates. J. Arid. Environ. 26, 47-51

Acorda JA, Acebedo MC and Maligaya RL. (2006) Ultrasonographic features of the liver and associated structures in the Philippine native goats (*Capra hircus*). Philipp J Vet Med 43 (1): 1-7.

Acorda JA, Ancheta MN, Detera MAJ, Cabrera LA and Maligaya RL. 2009a) Comparative ultrasound features and echo histograms of the spleen in female goats (*Capra hicus*), sheep (Ovis aries) and buffaloes (*Bubalus bubalis*). Philipp J Vet Anim Sci 35 (2): 135-146.

Acorda JA, Paloma JC, Cariaso WE and Cabrera LA. (2009b). Comparative ultrasound features of the liver, kidneys and spleen in female sheep (*Ovis aries*) at different ages. Philipp J Vet Med 46 (1): 26-36.

Acorda JA, Yamada H and Ghamsari SM. (1994) Evaluation of fatty infiltration of the liver in dairy cattle through digital analysis of hepatic ultrasonograms. Vet Radiol Ultrasound 35 (2):120-123.

Acorda JA, Yamada H and Ghamsari SM. (1995) Comparative evaluation of hydropic degeneration of liver in dairy cattle though biochemistry, ultrasonography and digital analysis. Vet Radiol Ultrasound 36 (4): 322-

Acorda JA and Alejandro VB. (2007) Ultrasonographic features of the liver and associated structures in female water buffaloes (*Bubalus bubalis*). Philipp J Vet Med 44 (2): 85-90.

Agarwal, S.P., Rai, A.K., Khanna, N.D. (1997) Induction of sexual activity in female camels during the nonbreeding season. Theriogenology 47, 591-600.

Ahmed, M.S.H. (1990) Some studies on the postpartum period in she camels. Camel News Letter, 7: 27

Alarcon V., Sumar J., Riera G.S., Foote W.C. (1990) Comparison of three methods of pregnancy diagnosis in Alpacas and Llamas. Theriogenology, 34: 1119-1127.

Allauigan JG. (2006) Ultrasonographic features of the kidneys in the Philippine water buffaloes (*Bubalus bubalis*) Undergraduate Thesis. University of the Philippines Los Baños.

Banerjee S.P., Vyas K.K., Pareek P.K., Deora K.S., Vyas U.K., (1981) Note on clinical diagnosis of early pregnancy in camel. Indian J. Anim. Sci. 51: 909-910.

Baruselli, P.S.; Mucciolo, R.G.; Vistin, J.A.; Viana, W.G.; Arruda, R.P; Maduriera, E.H.; Oliveira, C.A. and Molero-Filho, J.R. (1997). Ovarian follicular dynamics during the estrous cycle in buffalo. Theriogenology. 47: 1531-1547.

Bega, G.; Lev-Toaff, A.; Kuhlman, K.; Kurtz A,; Goldberg, B. and Wapner, R.(2001). Three dimensional ultrasonographic imaging in obstetrics: present and future. J Ultrasound Med., 20:391–408.

Bergfelt, D. R. Pierson RA and Ginther OJ (1989). Resurgence of the primary CL during Pregnancy in the mare. Animal Reproduction Science 21, 261-270.

Birnholz J.C., Farrel E.E. (1984) Ultrasound images of human fetal development. American Sci. 72: 608-613.

Blackwell R (1988) Safety of diagnostic ultrasound. In: Practical Ultrasound. Edited by Lerski RA. IRL Press, Washington DC. Pp 231-240.

Bobe G, Amin VR, Hippen AR, She P, Young JW and Beitz DC. (2008) Non-invasive detection of fatty liver in dairy cows by digital analysis of hepatic ultrasonograms. J Dairy Res 75: 84-89.

Bourke D.A., Adam C.L., Kyle C.E. (1992) Ultrasonography as an aid to controlled breeding in the llama (Lama glama). Vet. Rec., 130: 424-428.

Boyd J.S. (1995) Real-time diagnostic ultrasound in bovine reproduction. In: Goddard, P.J. ed., Veterinary Ultrasonography, Wallingford, U.K., CAB International, pp. 233-257.

Braun U (2003) Ultrasonography in gastrointestinal diseases in cattle. Vet J. 166 (2) 112-124.

Braun U, Gerber D and Sheffer U. (1992) Ultrasonography of the urinary tract of female sheep. Am J Vet Res 53 (10):1743-1739.

Braun U. (1990) Ultrasonographic examination of the liver in cows. Am J Vet Res 51 (10):1522-1526.

Braun U. (1991) Ultrasonographic examination of the right kidney in cows. Am J Vet Res 52 (12):1933-1939.

Braun U. (2009) Ultrasonography of the liver in cattle. Vet Clin Food Anim Prac 25:591-609.

Braun U and Gerber D. (1994) Influence of age, breed, and stage of pregnancy on hepatic ultrasonographic findings in cows. Am J Vet Res 55 (9):1201-1205.

Braun U and Hausammann K. (1992) Ultrasonographic examination of the liver in sheep. Am J Vet Res 53 (2): 198-202.

Braun U and Sicher D. (2005) Ultrasonography of the spleen in 50 healthy cows. Vet J 171: 513-518.

Burns PN, Waldroup L and Pinkney MN (1992) Limitation of ultrasound imaging measurements: In Atlas of ultrasound measurements edited by Goldberg BB and Kurtz AB. Year Book Medical Publishers, Inc, Chicago.

Bushong SC and Archer BR (1991) Diagnostic ultrasound: Physics, Biology and instrumentation. Mosby Year Book, ST. Louis. Pp 143-154.

Cartee RE (1980a) Diagnostic ultrasonography. Mod Vet Practice 61: 744-747.

Cartee RE (1980b) Ultrasonography: A new diagnostic technique for veterinary medicine. Vet Med / Small Anim clin 75: 1524-1533.

Cartee RE, Powe TA, Gray BW, Hudson RS, Kuhlers DL (1986) Ultasonographic evaluation of normal boar testicles. Amer J Vet Res 47:2543-2548.

Chandolia R.K., Nain R.S. and Verma S. K. (2001) Ultrasonographic studies on fetal development: Does a lamb learn rumination in the womb of mother?Biology of Reproduction (Suppl. 1) 64: 332 (Abstr. 576).

Chandolia RK, Jitendra K., Verma SK and Tayal R. (2002) Studies on comparative fetal activities of a goat kid and canine pups in utero: A goat kid ruminates in the womb of mother. Biology of Reproduction (Suppl. 1) 66:230 (Abstr. 327).

Chandolia RK, Pradeep SV, Jitender K and Tayal R (2005) Comparative ultrasonographic foetal studies in goat and dog. Indian Journal of Animal Sciences 75 (9) 1054 – 1056.

Chandolia RK. (2006) Ultrasonography in small ruminant reproduction. National convention of small ruminants at Avikanagar, Rajasthan, India

Chen, B.X. and Yuen, Z.X. (1979) Reproductive pattern of the Bactrian camel. In ; W.r.Cockrill (editor), The camel : an All purpose Animal, Scandinavian Institute of African studies, uppsala pp. 364-396.

Curan, S. (1992). Fetal sex determination in cattle and horses by ultrasonography. Theriogenology. 37: 17-21.

Curran S., Pierson R.A., Ginther O.J. (1986a). Ultrasonographic appearance of the bovine conceptus from day 10 through 20. J. American Vet. Med. Assoc., 189: 1289-1292.

Curran S., Pierson R.A., Ginther O.J., (1986b) Ultrasonographic appearance of the bovine conceptus from day 20 through 60. J. American Vet. Med. Assoc., 189: 1295-1302.

Deng, J. and Rodeck, C.H.(2004). New fetal cardiac imaging techniques. Prenat Diagn.,24:1092–1103

Djellouli M.S., Saint-Martin G. (1992) Productivity and economy of camel breeding in Tunisia. In: Proceedings of the First International Camel Conference, 2-6 Feb. 1982, Dubai, eds, Allen W.R., et al., R. and W. Publications (Newmarket) U.K., pp. 209-212.

Dorenberger V, Dornberger G and Eggstein M (1986) Volumetrie des Hodens mittels Real-time Sonographie. Ultraschall 7: 300-303.

Downey, D.B.; Fenster, A. and Williams, J.C. (2000) Clinical utility of three dimensional US. Radiographics. 20: 559-71.

Dyson, R. L.; Pretorius, D. H. and Budorick, N. E et al., (2000).Three- dimensional ultrasound in the evaluation of fetal anomalies, Ultrasound Obstet. Gynecol. 16:321–328.

El-Wishy, A.B. (1987) Reproduction in the female dromedary (*Camelus dromedarius*): A review. Anim. Reprod. Sci. 15: 273-297.

El-Wishy, A.B., Ghoneim, I.M. (1986) Breeding activity of the camel (*Camelus dromedarius*). Anim. Reprod. Sci. 11, 75-77.

Elias, E. (1990) Early weaning and post-partum conception in the one humped camel (Camelus dromedarius)- A review. In: G. Saint-Martin (editor), Proceedings of the Workshop "Is it possible to improve the reproductive performance of the camel?" CIRAD-EMVT, Paris, pp. 239-255.

Elias, E., Bedrak, E., Cohen, D. (1985) Induction of oestrus in the camel (*Camelus dromedarius*) during seasonal anoestrous. J. Reprod. Fert. 74, 519-525.

Elias, E., Bedrak, E. and Yagil, R. (1984) Peripheral blood levels of progesterone in female camels during various reproductive stages. Gen. Comp. Endocrinol., 53: 235-240.

England, B.G., Foote, W.C., Cardozo, A.G., Matthews, D.H. and Riera, S. (1971) Oestrous and mating behaviour in the llama (*Llama glama*). Anim. Behaviour, 19: 722-726.

Evans, J.O. and Powys, J.G. (1984). camel husbandry in Kenya: increasing the productivity of ranchland. In: W.R. Cockrill (Editor, The Camelid. An All Purpose Animal, Scandinavian Institute of African Studies pp. 347-357.

Fenster, A. and Downey, D.B. (2000) Three-dimensional ultrasound imaging. Annu Rev Biomed Eng. 2:457-475.

Fernandez, L.J.; Aguilar, A. and Pardi, S. (2004) Three-dimensional ultrasound in small parts: Is it just a nice picture? Ultrasound Quarterly.;20:119-25

Fish P (1990) Physics and Instrumentation of diagnostic Medical Ultrasound. John Wiley &Sons, New York.

Fissore, R.A.; Edmondson, A.J.; Pashen, R.L and BonDurant, R.H. (1986). The use of ultrasonography for the study of the bovine reproductive tract. II Nonpregnant, pregnant, and pathological conditions of the uterus. Anim. Reprod. Sci., 12: 167.

Floeck M. 2009. Ultrasonography of bovine urinary tract disorders. Vet Clin Food Anim Pract 25: 651-667.

Fowler D.G., Wilkins J.C. (1984) Diagnosis of pregnancy and number of fetuses in sheep by real-time ultrasound imaging. I. Effect of number of fetuses, stage of gestation, operator and breed of ewes on accuracy of diagnosis. Livest. Prod. Sci. 11: 437-450.

Frandson RD, Lee WW and Fails AD. (2003) Anatomy and Physiology of Farm Animals. Philadelphia: Lippincott Williams and Wilkins.

Gazitua F.J., Corradini C., Ferrando G., Raggi L.A., Victor H.P., (2001) Prediction of gestational age by ultrasonic fetometry in llama (*Lama glama*) and alpacas (*Lama pacos*). Ani. Reprod. Sci., 66: 81-92.

Ginther, O.J., Knopf, L. and Kastilic, J.P. 1989. Temporal association among ovarian events in cattle during oestrous cycle with two and three follicular waves. J. Reprod. Fert. 87:223-230

Ginther O.J., 1983. Mobility of the early equine conceptus. Theriogenology, 19: 603-609.

Ginther OJ (1995) Ultrasonic imaging and animal reproduction: Fundamentals: Book 1. Equiservices publishing, Wisonsin.

Goldstein, A. 1991. Physics of ultrasound. In Diagnostic Ultrasound -Vol. One. Morby year Book, London. Pp. 2-18

Goldstein A (1991) Physics of ultrasound. In: Diagnostic Ultrasound, Vol 1. Edited by Rumack CM,Wilson SR, Charbomeau JW. Mosby Year Book, St. Louis. Pp 2-18.

Goncalves, L. F.; Nien, J. K.; Espinoza, J.; Kusanovic, J. P.; Lee, W.;, Swope B, Soto,E.; Treadwell, M.C. and Romero, R. (2006). What does 2- dimensional imaging add to 3- and 4-dimensional obstetric ultrasonography?. J Ultrasound Med., 25(6):691-699.

Gonçalves, L. F.; Espinoza, J.; Kusanovic, J. P.; Lee, W.; Nien, J. K.; Forgas, J. S.; Mari,G.; Treadwell, M.C. and Romero, R. (2006) Applications of 2-Dimensional Matrix Array for 3- and 4-dimensional Examination of the Fetus. J Ultrasound Med., 25:745–755.

Griffin PG and Ginther OJ (1992) Research applictions of ultrasonography in reproductive biology. J Anim Sci. 70: 953-972.

Hildebrandt, T.B.; Drews, B.; Kurz, J.; Rollig, K.; Yang, S. and Goritz, F. (2007). 3D and 4D Ultrasonography: a new tool in veterinary research. BIR News., 1: 8–10.

Hildebrandt, T. B.; Drews, B.; Kurz, J.; Rollig, K.; Yang, S. and Goritz, F.(2009) Pregnancy monitoring in dogs and cats using 3D and 4D Ultrasonography. Reprod.Dom.Anim ., 4:123-125

Hsu, K.F.; Su, J.M. and Huang, S.C. et al. (2004) Three-dimensional power Doppler imaging of early-stage cervical cancer. Ultrasound Obstet Gynecol ;24:664-71

Janos, N. (1997) Three dimensional ultrasonogeraphy of the eyeball and orbit. Ophthalmology. 34:84-88.

Joshi, C.K., Vyas, K.K. and Pareek, P.K., 1978. Studies on the estrous cycle in Bikaneri camels. Indian J. of Anim. Sci., 48:141-145.

Jurkovic, D. (2002). Three-dimensional ultrasound in gynecology: a critical evaluation. Ultrasound. Obstet. Gyneco., 119:09–117.

Kahn, W. (1990). Sonographic imaging of the bovine fetus. Theriogenology. 33:385-396.

Kahn, W. (1994). Veterinary Reproductive Ultrasonography. Mosby-Wolfe, Kaufungen.

Khanna N.D., Tandon S.N., Rai A.K., 1990. Breeding parameters of Indian camels. Indian J. Anim. Sci., 60: 1347-1354.

Khurana, A, and Dahiya, N. (2004). 3D and 4D Ultrasound: A Text and Atlas. first ed., Anshan, U.K. Lazebnik, R. S. and Desser, T. S.(2007) Clinical3D ultrasound imaging: beyond obstetrical appliacations.SCME ultrasound

Kossoff G (1979) Analysis of focusing action of spherically curved transducer. Ultrasound Med Biol 5:359-368.

Kratochwil, A. (1992) Attempt at three-dimensional imaging in obstetrics.Ultraschall Med. 13: 183-6.

Kremkau FW (1984) Diagnostic ultrasound: Principles, Instrumentation, and Exercises. 2nd ed.Grune & Stratton, Inc. Toronto.

Kremkau FW (1991) Biological effects and safety. In: Diagnostic Ultrasound. Edited by Rumack CM, Wilson SR, Charboneau JW. Mosby Year Book, St. Louis.

Kumar J, Chandolia R.K., and Verma S. K. (2001) Ultrasonographic studies on fetal development in goats: activities of kid in utero. Biology of Reproduction (Suppl. 1) 64: 337 (Abstr. 589).

Kumar J, Chandolia RK and Verma SK (2005) Ultrasonographic detection of fetal mortality in a goat. Indian Vet. J.82 (8):894 – 895.

Kumar J, Chandolia RK and Verma SK (2005) Ultrasonographic imaging of early fetal development in Black-Bengal Goats. Indian Journal of Animal etuReproduction 26 (1): 39-42.

Lean I J, Abe N, Duggan S and Kingsford N (1992) with in between observer agreement in ultrasonic evaluation of bovine ovarian structures. Australian Veterinary Journal 69, 279-282.

Lee, H. J.; Choi, B. i.; Han, J. K.; Kim, A.Y.; Kim, K. W.; Park, S. H.; Jeong, J. Y. and Kang, J.W. (2002) Three-dimensional Ultrasonography Using the Minimum Transparent Mode in Obstructive Biliary Diseases. J. Ultrasound. Med., 21: 443–453.

Lee, W. (2003).3D fetal ultrasonography. Clin. Obstet. Gynecol., 46: 850– 867.

Lees, W. (2001) Ultrasound imaging in three and four dimensions. Seminars in Ultrasound, CT & MR.;22:85-105.

Leung, K.Y, Ngai, C.S.W, and Chan, B.C. et al. (2005) Three-dimensional extended imaging: a new display modality for three-dimensional ultrasound examination.Ultrasound Obstet Gynecol; 26:244-51

Little TV and Woods GL (1987) Ultrasonography of accessory sex glands in the stallion. J Reprod Fertil., Suppl 35: 87-94.

Love CC, Garcia MC, Riera FR, Kenney RM (1990) Evaluation of measures taken by ultrasounography and calipers to estimate testicular volume and predict daily sperm output in the stallion. J Reproduction Fertil Suppl., 44: 99-105.

Malmgren L and Sussenmilch BI (1992) Ultrasonography as a diagnostic tool in a stallion with seminal vesiculitis: Acase report. Theriogenology: 37: 935-938.

Marie M., Anouassi A., 1986. Mating induced luteinizing hormone surge and ovulation in the she camel (Camelus dromedarius). Biol. Reprod., 35:792-798

Martinat-Botte F., Renaud G., Madec F., Costiou P.,Terqoi M. 1998. Echographie et reproduction chez les triee. INRA/Hoechst Vet, Paris, p. 104.

Maslak SM (1985) Computed sonography. In : Ultrasound annual edited by RC Sanders & M.c. Hill.

Matharu, B.S.,1966. Camel care. Indian Farm., 16, 19-22.

Maymon, R.; Herman, A. and Ariely, S. et al. (2000) Three-dimensional vaginal sonography in obstetrics and gynaecology. Hum Reprod Update; 6:475-84.

McDicken WN (1991a) Diagnostic Ultrasonics: Principles and Use of Instruments. 3rd Ed. Churchill Livingstone, New York City.

McDicken WN (1991b) Quality assurance for echo imaging. In: Diagnostic Ultrasonics: Principles and Use of Instruments. 3 rd Ed. Churchill Livingstone, New York. Pp 323-328.

Mc Donald L.E. (1980) Veterinary Endocrinology and Reproduction. Lea and Febiger, Philadelphia.

Mialot J.P., Villemain L., 1994. Ultrasonic diagnosis of gestation in the llama and alpaca. Rec. Med. Vet., 170: 23-27.

Miller MW and Ziskin MC (1989) Biological consequences of hyperthermia. Ultrasound Med Biol 15:707.

Millner R, Rosenfeld E and Cobet U (editors) (1983) Ultrasound interactions in biology and medicine. Plenum press New York.

Minoia, P., Moslah, M., Lacalandra, G.M., Khorchani, T., Zarrilli, A., 1992. Induction of estrus and management of reproduction in the female dromedary camel. In Proceedings of the First International Camel Conference, Dubai, U.A.E., February 2-6 1992, pp. 119-127.

MiskinM, Bukspan M and Bain J (1977) Ultrasonographic examination of scrotal masses. J Urol 117:185-188.

Musa, B.E. and Makawi, S.A., 1989. Involution of the uterus and the first post partum heat in the camel (*Camelus dromedarius*). Camel News Letter, 5: 7-8.

Musa B.E., Abusineina M.E., 1978. Clinical pregnancy diagnosis in the camel and a comparison with bovine pregnancy. Vet. Rec., 102: 7-10.

Neeraj Arora (2013) Ultrasonographic and haemato - biochemical studies of omasum and abomasum disorders in buffaloes. Ph.D. Thesis, Submitted to Lala LAjpat

Nyland TG and Matoon JS. 1995. Veterinary Diagnostic Ultrasound. Philadelphia: WB Saunders Company.

Paraguez V.H., Cortez S., Gazitera F.J., Ferrando G., Macnicen V., Raggi L.A., 1997. Early pregnancy diagnosis in alpaca (*Lama pacos*) and llama (Lama glama) by ultrasound. Ani. Reprod. Sci., 47: 113-121

Pawshe C.H., Apparao K.B.C., Totey S.M., 1994. Ultrasonographic imaging to monitor early pregnancy and embryonic development in buffalo (*Bubalus bubalis*). Theriogenology, 41: 697-709.

Pechman RD, Eilts BE (1987) B-mode ultrasonography of the bull testicle. Theriogenology 27: 431-441.

Pierson, R. A. and O.J. Ginther (1984). Ultrasonography of the bovine ovary. Theriogenology 21, 495-504.

Pierson RA, Kastelic JP and Ginther OJ (1988) Basic principles and techniques for transrectal ultrasonography in cattle and horses. Theriogenology 29: 3-20.

Purandare, C. N. (2006) 3D – 4D Ultrasound in Obstetrics and Gynecology J Obstet Gynecol India.. 56(1):22-24

Rankin, R.N.; Fenster, A and Downey, D.B, et al. (1993)Three dimensional sonographic reconstruction: techniques and diagnostic application.Am J Roentgenol; 161: 695-702.

Rantanen NW and Ewing RL (1981) Principles of ultrasound application in animals. Vet Radiol 22:196-20

Reece WO. 1991. Physiology of Domestic Animals. Pennsylvania: Lea and Febiger.

Rifkin MD (1992) Measurements of scrotal contents. In: Atlas of ultrasound measurements edited by B.B. Goldberg and A.B.Kurtz. Year book Medical Publishers, Inc Chicago.

Rifkin MD, Kurtz AB, Choi HY and Goldberg BB (1983) Endoscopic ultrasonic evaluation of the prostate using a transrectal probe: prospective evaluation and acoustic characterization. Radiology 149, 265-271.

Roberts, SJ (1986) Veterinary Obstetrics and Genital Diseases (Theriogenology). WoodstockVT: S J Roberts, Publisher: 411-412.

Romero JM, Finger PT. Three dimensional ultrasonogeraphy of measurement of choroidal melanoma. Am J Ophthalmol 1998; 126:842-846.

Russel AJF (1989) The application of real time ultrasonic scanning in commercial sheep, goats and cattle production enterprises. In: Diagnostic Ultrasound and Animal Reproduction. Edited by Taverne MAM,Willemse AH. Kluwer Academic Publishers, Boston pp 73-87.

Sampler WF, Gotlesman JE , Skinner DG et al (1978) Gray scale ultrasound of the scrotum. Radiology 127: 225-228.

Sarti DA and Sampler WF (1980) Diagnostic Ultrasound; Text and Cases. G. K. Hall and Co., Boston.

Sghiri, A., Driancourt, M.A., 1999. Seasonal effects on fertility and ovarian follicular growth and maturation in camels (*Camelus dromedarius*). Anim. Reprod. Sci., 55: 223-237.

Sharma, R.K. (2003).Clinical and research application of ultrasound in reproduction.In: Artificial breeding and reproduction management in buffaloes. Editors: Inderjeet Singh, P.S.Yadav and R.K.Sharma. Summer school held at CIRB, Hisar from June 10- 30, 2003. pp 113-118.

Sharma, R.K.;Singh, J.K.; Singh, Inderjeet and Khanna, S. (2004). Development of persistent ovarian follicles in prepubertal Murrah buffalo heifers following exogenous GnRH. In: Proc. of National Symposium on 'Livestock Biodiversity vis a vis Resource exploitation: An introspection'. Held at NBAGR Karnal 11-12th Feb., 2004. pp.198.

Sharma , S.S. and Vyas, K.K., 1972. Involution of the uterus and vulva in the camels. Ceylon Vet. J.: 20, 9-10.

Shirley IM, Blackwell RJ, Geoffrey C, Farmen DJ and Ricary FR (1978) A user's guide to diagnostic ultrasound. University park press, Baltimore.

Skidmore, J.A., Billah, M. and Allen, W.R. (1997). The ovarian follicular wave pattern and control of ovulation in the mated and non-mated dromedary camel (*Camelus dromedarius*). J. Camel Prac. Res. 4: 193-197.

Skidmore J.A., Billah M., Allen W.R., (1996a). The ovarian follicular wave pattern and induction of ovulation in the mated and non-mated one- humped camel (Camelus dromedarius). J Reprod. Fert., 106:185:192

Skidmore J.A., Billah M., Allen W.R., 1992. Ultrasonographic and video- endoscopic monitoring of early fetal development in the dromedary camel. In: Proceedings of the First International Camel Conference, 2-6 Feb. 1982, Dubai, eds, Allen W.R., et al., R. and W. Publications (NEWMARKET) U.K., pp.193-201.

Slapa, R. Z.; Jakubowski, W. S.; Srzednicka, J. S. and Szopinski, K.T. (2011) Advantages and is advantages of 3D ultrasound of thyroid nodules including thin slice volume Rendering. Thyroid Research. 4:1

Spaulding KA (1992) Use of an ultrasound phantom. Vet Radiol and Ultrasound 33:199-200.

Starke A, Huadum A, Weijers G, Herzog K, Wohlsein P, Beyerbach M, De Korte L, Thijssen JM and Rehage J. 2010. Noninvasive detection of hepatic lipidosis in dairy cows with calibrated ultrasonographic image analysis. J Dairy Sci 93: 2952-2965.

Tarantal AF and O'Brien WD Jr (1994) Discussion of Ultrasonic Safety related to obstertics. In Diagnostic Ultrasound applied to obstetrics and gynaecology, edited by R.E. Sabbagha, J.B. Lippincot co., Philadelphia.

Taverne MAM andWillemse AH (editors) (1989) Diagnostic Ultrasound and Animal Reproduction. Kluwer Academic Publishers, Boston.

Tibary A., Anouassi A., 1996. Ultrasonographic changes of the reproductive tract in the female camel (Camelus dromedarius) during the follicular cycle and pregnancy. J. Camel Prac. Res.3: 71-90.

Tibary A., Anouassi A., 1997. Theriogenology in Camelidae: Anatomy, Physiology, pathology and artificial breeding. UAE, Ministry of culture and information. First edition. Chapter VII: Breeding soundness examination of the female camelidae, pp. 281-286.

Timor-Tritsch, I.E.; Han, E. and Platt, L.D. (2002) Three-dimensional ultrasound experience in obstetrics. Curr Opin Obstet Gynecol.14:569- 75.

Tinson A.H., Mckinnon A.O., 1992. Ultrasonography of the reproductive tract of the female dromedary camel. In: Proceedings of the First International Camel Conference, 2-6 Feb. 1982, Dubai, eds, Allen W.R., et al., R. and W. Publications (Newmarket) U.K., pp. 129-135.

Tonni, G.; Centini, G. and Rosignoli, L. (2005). Prenatal screening for fetal face and clefting in aprospective study on low-risk population: can 3- and 4-dimensional ultrasound enhancevisualization and detection rate?. Oral Surg. Oral Med. Oral Pathol. Oral Radiol. Endod., 100: 420–426.

Vosough,D.; Masoudifard,M.; Veshkini.A; Vajhi, A. R.; Sorouri,S (2007) Three Dimensional Ultrasonography of the Eye and Measurement of Optical Nerve Sheet Diameter in Dog. Iranian j. veterinary surgery2 (2):73-77.

Vyas, S., Rai, A.K., Sahani, M.S. and Khanna, N.D. (2004) Use of real time ultrasonography for control of follicular activity and pregnancy diagnosis in the one humped camel (Camelus dromedarius) during non- breeding season. Animal Reproduction Science 84 (1-2): 229-233.

Vyas, S., Sahani M.S., 2000. Real time ultrasonography of ovaries and breeding of the one humped camel (*Camelus dromedarius*) during the early postpartum period. Anim. Reprod. Sci. 59: 179-184.

Wani G.M., 1987. Ultrasonographic pregnancy diagnosis in sheep and goats. World Rev. Anim. Prod. XVII 4: 43-48.

Warkany J (1986) Teratogen update: Hyperthermia. Teratology 33:365

Weber JA, Hilt CJ and Woods GL (1988) Ultrasound appearance of bull accessory sex glands. Theriogenology 29: 1347-1355.

Weber JA and Woods GL (1991) a technique for transrectal ultrasonography of stallions during ejaculation. Theriogenology 36: 831-837.

Wicks JD and Howe KS (1983) Fundamentals of Ultrasonic Technique. Year Book Medical Publishers, Inc. Chicago.

Williamson, P. and Payne, W.J., 1978. An introduction to animal husbandry in the tropics, 3rd edn. Longman, London, pp.484-498.

Wilson R.T., 1989. Reproductive performance of the one humped camel. The empirical base. Revue E'lev. Med. vet. Pays trop., 42: 117-125.

Winsberg F, Cooperberg PL (1982) Real time ultrasonography. Churchill Livingstone, Newyork city.

Won, H. J.; Han, J. K.; Do, K. H.; Le, K. H.; Kim, K.W.; Kim, S. H.;Yoon, C. J.; Kim, YJ,; Park, C. M, and Choi, B. I. (2003). Value of Four- dimensional Ultrasonography in Ultrasonographically Guided Biopsy of Hepatic Masses. J. Ultrasound. Med .,22:215–220.

Yagil, R. and Etzion,Z., 1984. Enhanced reproduction in camels (*Camelus dromedarius*). Comp. Biochem. Physiol., 79A: 201-204.

Yamaga Y and Too K. 1984. Diagnostic ultrasound imaging in domestic animals: fundamental studies on abdominal organs and fetuses. Jpn J Vet Sci 46 (2): 203-212.

Yasin, S.A. and Wahid, A., 1957. Pakistan camels- a preliminary survey. Agric. Pak., 8: 289-297

Zagzebski J A (1983) Images and artifacts. In: Textbook of Diagnostic Ultrasound. Edited by Hagen-Ansert S.C.V. Mosby Co., St. Louis.

Zagzebski J A (1994) Physics and Instrumentation. In: Diagnostic Ultrasound Applied to Obstetrics and Gynaecology, edited by Rudy E Sabbagha, I.B. Lippincott company, Philadelphia. uthors Introduction:

Index